1995

A PRACTICAL PHILOSOPHY FOR
THE LIFE SCIENCES

SUNY Series in Philosophy and Biology
David Edward Shaner, editor, Furman University

A PRACTICAL PHILOSOPHY FOR THE LIFE SCIENCES

WIM J. VAN DER STEEN

State University of New York Press

Published by
State University of New York Press, Albany

© 1993 State University of New York

Printed in the United State of America

For information, address States University of New York Press,
State University Plaza, Albany, N.Y., 12246

Production by Marilyn P. Semerad
Marketing by Nancy Farrell

Library of Congress Cataloging-in-Publication Data
Steen, Wim J. van der, 1940-
    A practical philosophy for the life sciences / Wim J. van der
Steen.
        p.   cm. — (SUNY series in philosophy and biology)
    Includes bibliographical references (p.   ) and index.
    ISBN 0-7914-1615-1 (hardcover). — ISBN 0-7914-1616-X (paper)
    1. Life sciences—Philosophy.   2. Logic.   3. Science—Philosophy.
I. Title.   II. Series.
QH331.S784   1993
574′ .01—dc20   92-27319
                                                                CIP

        10   9   8   7   6   5   4   3   2   1

for Dick Burian
and Anne,
who made me taste
the milk of human kindness

Contents

CONTENTS

Preface

Styles in this book range from conversational to technical. The conversational style develops into a general view in chapter 10. That chapter can be read in advance as a preview of major themes.

The book is primarily a textbook. Hence dull technicalities are unavoidable. The chapter on formal logic is the most technical one. The composition of the text practically allows you to do without it. However, I would not advise you to skip formal logic in case you aim at analytical skills.

The text presupposes some familiarity with biology or medicine. Prior knowledge of philosophy or ethics is not required. Biology, medicine, philosophy and ethics are presented in an integrative fashion which is new. The book may come in useful among students and professionals in any of these areas.

The text has four major functions.
- Philosophy is made practical for the life sciences. This function has the main emphasis.
- Empirical science is infused into the philosophy of science and into ethics.
- New avenues are implicated for research in the philosophy of biology and the philosophy of medicine.
- Limitations of science, philosophy and ethics are uncovered. Hence the emphasis on *context-dependence* throughout the book.

Most chapters contain examples and exercises focussing on live science. Answers to exercises are appended to the main text.

Acknowledgments

Let this be a chronology.

First phase, a couple of years ago. The bad start. I was used to writing books, in Dutch mostly, in close cooperation with publishers. Discussion of drafts, exchange of info and wishes, drastic revisions and all that. I should have told David Shaner, the series editor whose help has been invaluable in all phases. I didn't, with results predictable. Devastating comments by referees. Meanwhile colleagues this side of the ocean provided extensive comments. Peter Sloep, Bart Voorzanger and Cor van der Weele helped me shape the second draft.

Second phase. The second draft was sent to colleagues in the US and Canada. Comments were kindly provided by Dick Burian, David Hull, Susan Oyama, Ken Schaffner, Sergio Sismondo, Elliot Sober, Paul Thompson and Pat Williams. Amidst a positive overall response, I received gems of constructive criticism. Specifically I owe much to extensive comments from Dick, David, Ken and Sergio. My new colleague Ad van Dommelen also gave helpful comments in this phase.

Third phase. I think the job is done. US criticism is accounted for in the text. Enter Cor van der Weele. Wim, I don't know why these US people are as positive as you say they are. Your revision is indigestible. That's how I remember her words. Maybe I exaggerate. Cor amplified this by arguments, succinct but incisive. Convincing she was, as I realized when the delayed shock came. Back to the grindmill. That's when I became a burden for others. Gratefully I thank Mineke for ignoring, with unparalleled wisdom, the capriciousness around my book writing. Three colleagues, Ad van Dommelen, Annelies Stolp and Cor van der Weele helped me prepare the final version by a rare combination of openness, sincerity and effective assistance. Bart Voorzanger kindly took care of remaining correction work.

Thea Laan belongs to all phases. She was the heart of the technical assistance, highly competent and always friendly.

CHAPTER 1

Introduction

In the last two decades or so, philosophy of biology and philosophy of medicine, here collectively called philosophy of the life sciences, have come of age. Many books and articles are now published each year in this area. The emphasis is on advanced level work. Hence this textbook, which may fill up the lacuna.

In philosophy controversies are more prominent than in science. Therefore writing a textbook on applied philosophy is a hard job. Of course one can concentrate on elementary logic and methodology which is relatively uncontroversial in philosophical circles. I will be doing that, but it is not enough. If you want to evaluate scientific theories and explanations with philosophical tools, you will have to address philosophical views on the subject. You would soon find out that there is no consensus in philosophy about the nature of theories and explanations.

I have tried to come to grips with this problem by doing philosophy in a style which is more common in science than in philosophy. Philosophers typically aim at generally valid models and theories about science. Scientists also cherish generality, but they tend to be more modest. If a model they have developed is not generally valid they may duly note exceptions and apply it to a restricted domain. I will deal with philosophical models in a similar fashion. They can be quite useful in some contexts even if they are apparently invalidated by counter-examples.

Some of the philosophical questions and distinctions which are central in this book are best introduced by an example. Let's briefly consider the subject of aging.

To some extent man is an exception among the animals. Some biologists regard the prevalence of aging in man as an important example. Many people *in our culture* reach old age and so come to experience the process of aging. Animals—in the wild—seldom do. Neither did people in prehistoric times.

The thesis that aging is rare in nature, if true, is important for theories of aging. Many theories are alleged to explain aging (see Warner *et al.*, 1987). There is no clear winner yet. I will briefly describe one theory, which some have declared a loser on the ground that aging is rare. The theory says that aging occurs because it is genetically determined, or programmed. The idea behind it is that development is controlled by a genetic program. Aging should be controlled in the same way because it belongs to a process of continuous development.

Hayflick (1987) has mounted the criticism that genetically programmed aging could not have evolved through natural selection. There have not been enough aged organisms for selection to work on. From this he concludes that aging cannot be genetically programmed. He apparently presupposes, without explicitly saying so, that all genetic programs are the result of natural selection.

Suppose we set out to evaluate Hayflick's line of reasoning. What questions should we ask about it? In the first place there are *questions concerning* the *truth or falsity* of statements representing evidence. Is it true that aging is rare in nature? Is it true that genetic programs are the result of natural selection? These questions must ultimately be answered by an appeal to facts.

'Ultimately' indeed. We should not confront these statements with facts unless we understand central *concepts*. Is the meaning of 'aging', 'genetic program' and 'natural selection' clear? I will indicate by comments on one concept that such *questions about meaning* are sometimes crucially important. 'Genetic determination', a concept embedded in the more complex concept of 'genetic program', will do.

Let's try a *definition*. 'A feature is genetically determined if genetic factors are among the causes of the feature.' Would this be an appropriate definition? No, it wouldn't, however plausible it may look at first sight. It is a plain truth of biology that *all* features of organisms result from genetic *and* environmental influences. If we adopt the definition we should say that all features are genetically determined. Those who distinguish genetically determined features from environmentally determined ones must have something different in mind.

The following definition will not do either. 'A feature is genetically determined if it is caused exclusively by genetic factors.' On this definition, there would not be *any* genetically determined features.

On the first definition, the thesis that aging is genetically determined is a trivial truth which makes a confrontation with evidence superfluous. On the second definition we get a thesis which is quite obviously false. How then should we define 'genetic determination' to get out of this curious situation?

The solution is that *we should not attribute genetic (or environmental) determination to features of individual organisms. Instead, we should consider differences between organisms.* The following definition is on the right track. The difference in feature *F* between organism *x* and organism *y* is genetically determined if the difference in *F* is due to a genetic difference between *x* and *y*. The unqualified statement that some feature is genetically determined obviously means that differences between organisms in this feature are always due, exclusively so, to a genetic difference. Few features are *genetically determined* in this *strong sense*. Eye color in adult humans is an example.

Conceptual analyses can have important consequences for the evaluation of evidence. What kind of *evidence* would we need for the thesis that aging is genetically determined? Perhaps we are able to demonstrate that a particular gene plays a causal role in aging in a particular organism. However, this would not be evidence in support of the thesis. The comments above show that we rather need information on *differences* between organisms.

Currently available evidence suggests that aging is not genetically determined in a strong sense. It is probable, though, that there is *genetic determination* in the *weaker sense* that *some* differences in aging are due to genetic differences or partly so. Concerning a genetic program for aging I will not hazard a conclusion. 'Genetic program' is a complex metaphor which involves much more than genetic determination alone. It is often used in biological texts without any clarification of its meaning. That's why I wouldn't know how to recognize evidence pointing to a genetic program.

Suppose we would come to the conclusion that the statements put forward by Hayflick are clear enough, and that they are likely to be true. That obviously would not suffice to accept his argument. In addition we will have to ask *questions about validity*. An *argument* is valid if its conclusion is necessarily true *if* the statements which are meant to support the conclusion—the premises—are true. We can easily see that Hayflick's argument is valid if we reconstruct it as follows. All genetic programs have evolved through natural selection. If there is a genetic program for aging, then it is not the case that all genetic programs have evolved through natural selection (this particular program being an exception). Therefore, there is no genetic program for aging.

If we use the symbol '*A*' for the statement 'All genetic programs have evolved through natural selection', and '*B*' for 'There is a genetic program for aging', Hayflick's argument takes the following form. '*A*; if *B*, then not-*A*; therefore: not-*B*.'

This is clearly a valid argument. Notice that *validity* is wholly *determined by form.* Whatever statements we care to substitute for '*A*' and '*B*', we will always get a valid argument.

It goes without saying that the premises of Hayflick's argument must themselves be based on arguments. (He does not provide arguments in an explicit form.) Thus the thesis that all genetic programs have evolved through natural selection is assumed to follow from premises that belong to evolutionary biology. We would have to analyse implications of evolutionary theory to reach a verdict on Hayflick's view (for more information see Rose, 1991, who endorses an evolutionary view of aging).

The discipline of *logic*, broadly defined, is concerned with the analysis of concepts and arguments forms. My comments above indicate that applications of logic are indispensable in scientific work.

Questions concerning the logic of science are relatively concrete and specific. In the evaluation of scientific reasoning we should also ask more general questions that belong to *methodology*, a branch of *philosophy of science*. Scientists elaborate *theories* by *testing hypotheses*. The theories are used to *explain* and *predict* phenomena. Methodology characterizes the nature of theories, tests, and so forth, in general terms, and formulates criteria they must satisfy. Methodology, like logic, is indispensable in science.

Logical criteria such as validity are relatively straightforward. *Methodological criteria* are more elusive. By way of an example, I will briefly consider the nature of *scientific theories* and the criteria they must satisfy. In physics, theories typically consist of laws of nature, highly general statements which are well-confirmed. Not so in biology. There are many theories of aging, but to my knowledge no law of aging has ever been formulated.

If we would take the view that scientific theories must satisfy the methodological criterion of *generality*, biology would compare poorly with physics. It would have few adequate theories if any. I regard this as an unreasonable view. My conclusion is rather that *the category of theories is heterogeneous.* Some theories are general, others are not.

Many methodological criteria for theory assessment have been considered in the literature: clarity, empirical content, realism, testability and confirmation by tests, consistency, coherence, generality, simplicity, explanatory power and predictive power. It should be obvious that *theories will seldom satisfy all* these *criteria* at the same time.

For example, generality may be at odds with realism. Should one opt for a theory which is general, or for one which is realistic? That is a question without a general answer. Theories about drug effects had better be realistic if we want to use them for calculating dosages of drugs which patients need. Such theories will not be general since they must deal with highly specific situations. In contrast to this, those who want to understand the spread of epidemics may well be content with general mathematical theories which disregard many factors with minor effects. More realistic epidemiological theories tend to be mathematically untractable.

The importance of methodological criteria obviously depends on the context, on the purposes we have.

The criteria of *clarity and empirical content are important* because they are presupposed by various other criteria. Therefore they will get much attention in this book. I guess I need not clarify clarity.

Empirical content is an elusive notion. A statement has empirical content if its truth or falsity depends only on the way it deals with *facts*. The term 'fact' in this connection must not be taken in the broad sense of ordinary language. We normally use the term 'fact' for many different things. It is a fact that $2 + 2 = 4$, that murder is bad, that I am now writing a book, that all metals expand when heated, that all bachelors are unmarried, and so forth. In science and philosophy the concept of fact mostly has a narrower meaning. It stands for things covered by statements that describe situations or events at a particular time and a particular place.

The statement that I am now (July 5, 1992, 1 a.m. local time) writing a book is true because it expresses a fact. Thus it has empirical content. The statement that I am not writing a book now likewise has empirical content. It is false because what it expresses is at odds with a fact—the same one. The general statement that all metals expand when heated describes a multitude of facts rather than a single one. It obviously has empirical content.

The statement that $2 + 2 = 4$ is different. Its truth does not depend on facts expressed by it; indeed it does not express any facts. Truth in this case is determined by linguistic 'facts' concerning the meaning of concepts ('2', '+', '=', '4'). So the statement has no empirical content.

Philosophers call statements with empirical content synthetic statements. I will use a more easy-sounding terminology and simply call them *empirical statements*.

Statements with truth or falsity depending on meanings are called analytic statements by philosophers. In addition to this, statements are also called analytic when their truth or falsity depends on form. 'I am a man or I am not a man' is an example. This statement is true but its truth depends neither on facts nor on meanings. Its *form* is '*p* or not-*p*'. *Any* statement with this form must be true. Likewise the form '*p* and not-*p*' will always yield false statements.

The truth or falsity of analytic statements can be determined by logical analysis, that is an analysis of meanings or of logical form. Because the concept of analyticity is seldom used outside logic and philosophy, I will speak of *logical statements* instead of analytic statements. Notice that I am using this expression for two categories of statements, those which are true or false in virtue of meanings, and those which are true or false in virtue of logical form.

Various philosophers, most notably Quine (1953, 1960), have rejected the distinction of logical and empirical statements since there is no sharp boundary. I would argue that the distinction is quite useful if we are willing to take the *context* into account. Statements take on meaning in the context of a particular *discourse*. Hence it is possible for a statement to be logical in one discourse and empirical in a different discourse. 'All bachelors are unmarried' is a logical statement if the term 'bachelor' is used for unmarried persons. In a different discourse, 'bachelor' may stand for persons who do not live with a partner. 'All bachelors are unmarried' would be a false empirical statement in that discourse, since some persons who do not live with a partner are in fact formally married.

Empirical statements and logical statements are *cognitive statements*, statements which are true or false. In addition to this there are *non-cognitive statements*, which cannot be true or false. Commands ('Shut the door')—if looked upon as statements—are an obvious example. What to think of 'Murder is bad'? It has the ring of a true statement. However, it would not be easy to specify what truth depends on in this case. On second thought we may come to doubt whether the statement can be true or false. In philosophy opinion on this is divided.

'Murder is bad' belongs to the category of *normative statements*. Statements in this category express values ('Murder is bad') or norms ('Murder must be punished'). I will stay on the safe side and refrain from applying the concepts of truth or falsity to normative statements. Instead I will call them *acceptable* or *unacceptable*.

The first few chapters of this book introduce the logic of concepts and arguments. Subsequently I consider topics such as hypothesis testing and explanation, which are more obviously in the domain of philosophy of science. In chapters *9* and *10* the focus is on relations between science and the domain of the normative. These chapters put science, philosophy and ethics in a broader perspective. Throughout the book, the emphasis is on applications in the life sciences.

CHAPTER 2

Concepts

2.1. The Logical and The Empirical

Concepts are the subject of this chapter. The subject is difficult, don't underestimate it. I know a sophisticated philosophical book which is one long argument designed to show that philosophical books produced over the centuries have not provided us with a coherent theory about the meaning of the concept of being a concept. If you don't understand the last sentence, don't worry. I'm not going to bother you with that kind of philosophy.

I will not preview problems in this chapter. My hunch is that it wouldn't convey much. Let's plunge into live science instead. Here is a specimen.

> The term 'homeostasis' is used to refer to the ability of the body (*a*) to maintain its essential parameters nearly constant and (*b*) to take corrective action to return them to normal following a disturbance. The expression 'homeostatic behaviour' can assume one of two closely related, but slightly different, meanings, reflecting (*a*) and (*b*) above ... . Let us consider drinking. An animal might drink in response to the loss of fluid from its body. On presentation of water, the loss provokes drinking until such a time as it is corrected. This behaviour would be termed 'homeostatic' by the criterion of the end that it serves, that of maintaining the constancy of body fluids. By an alternative criterion, that of the *mechanism* underlying behaviour, it would also be termed homeostatic. ... Suppose though, that on another occasion the animal drinks, not in response to a deficit, but in anticipation of a future deficit. Some authors would not describe this as homeostatic, since it is not driven by a displacement from the homeostatic norm. Other authors, emphasizing the end served, would term it homeostatic ... .
>
> By the criterion of behaviour that serves a regulatory end, ... is it ever possible to envisage non-homeostatic drinking or feeding? It is difficult, since clearly evolution will not favour such a mechanism. However, in the laboratory, it is possible to show such behaviour ... (Toates, 1986, pp. 35-36).

9

This passage from a book by Toates, the ethologist, confronts us with a characteristic feature of science. It is concerned with *empirical matters*. In addition to this, at a different level, Toates discusses *logical matters* concerning the use of language. To do science, or to understand it, we need to know about logic. This is an obvious truth, one which is by no means trivial. In science problems of logic can be as hard as empirical problems.

Toates's line of reasoning illustrates that logical matters affect the evaluation of empirical theses. The thesis that drinking and feeding are homeostatic is presumably true under one interpretation of 'homeostatic'. Under a different interpretation it is probably false. Hence it is wise to *consider questions of meaning*—which belong to logic, broadly conceived—*before addressing questions of truth.*

The logic of scientific language will be an important theme in this book. Language can be studied at three levels, *words, sentences and arguments*. The present chapter deals with words, or *terms*, the simplest building blocks of language, and *concepts*, the ideas expressed by words. It is important to distinguish between discourse about things and situations, and discourse about language. A common convention to mark the distinction, which I will adopt, is the use of *quotation marks to indicate linguistic entities*. Thus a horse is an animal and 'horse' is a word. 'Horse' will also indicate the idea behind the word 'horse', the concept of horse. Likewise 'Horses are animals' indicates a *sentence*, or the idea behind it—the *statement* or *proposition* expressed by it.

The proper use of words and concepts is governed by *methodological criteria or principles*. First and foremost it is advisable to aim at clarity. *According to the criterion of clarity vagueness and ambiguity are to be avoided, though not at all costs.*
    A word or concept is *vague* if boundaries of its domain of application are not sharp. Words and concepts for colours in ordinary discourse are a case in point. In looking at a rainbow it is hard for us to tell where green ends and blue begins. The words 'green' and 'blue' are vague. So are the underlying concepts.
    Ambiguous words or concepts have more than one meaning. 'Explain' is an example. It has numerous meanings in daily life as well as in science and philosophy.

Vague or ambiguous words need not be associated with vague or ambiguous concepts. Consider the following two cases. (i) If you don't get a point I am making, you may ask me to *explain* what I mean. (ii) If you are baffled by a phenomenon, you may ask me to *explain* why it occurred.
    The *word* 'explain' is indeed ambiguous, it means very different things in the two cases. I guess that I need not explain (!) this. Now one could argue that there is no such thing as *one concept* 'explain'. On this view the word 'explain' is used for different concepts which might be unambiguous individually.

In this book I often disregard distinctions of this kind to keep the text accessible. Unless the occasion demands it I do not distinguish words and concepts. Thus I will freely use 'concept' for concepts together with words.

The criterion of clarity as I formulated it contains a qualifying phrase: vagueness and ambiguity are to be avoided, *though not at all costs.* This phrase is appropriate since the importance of clarity is *context-dependent.* If your purpose is to formulate a rigorous theory which concludes an episode of extensive research, you had better end up with clear concepts. In ongoing research the situation can be quite different, for example, when uncertainties call for vague formulations. Thus vagueness is at times a virtue rather than a vice. Likewise for ambiguity. Researchers who disagree about facts will often use concepts in different ways, rightly so. After all concepts are designed to 'fit' the facts as researchers see them. However, I would take the stance that if vagueness and ambiguity cannot be avoided, they must be brought out into the open whenever that is possible, to prevent unnecessary confusion.

The following example is a starter which indicates that vagueness is easily overlooked in scientific writings.

• *Example 1.* 'Abundance' is a term used by ecologists to describe 'commonness' and 'rareness' of organisms. If you come across it and your mood is not geared to analysis, you will often feel—as I do—that the meaning of the concept is totally clear. So it may be, *in the context of the text you are reading with all kinds of presuppositions which go unmentioned.* The concept of abundance will not cause trouble in many cases, but troublesome cases do occur. A natural way to express abundance is by numbers of organisms. Thus the thesis that organism *A* is abundant in some area will mean that there are many individuals of *A* in the area. However, this makes sense only if the distribution of *A* is more or less even. If *A* locally has numerous individuals whereas it is absent elsewhere, the statement that it is abundant in the area is inappropriate. Likewise the statement that elephants are less abundant in an area than bacteria is awkward. 'Abundance' is apparently a vague concept since it is impossible to be very specific about the domain to which it applies. In many cases it is wise to replace it by more precise concepts. •

Achieving clarity—whenever desirable and possible—*is anything but easy since vagueness and ambiguity are pervasive in ordinary language.* Even seemingly innocuous words such as the verb 'to be' can be a source of trouble. In some cases this verb represents a logical connection. In other cases the connection is empirical. The subject of ambiguity thus brings us back to the distinction of *logical statements* and *empirical statements* which was introduced in the previous chapter. Examples 2-4 illustrate various links which the distinction can have with clarity, vagueness and ambiguity. It is not my intention to provide you with a neat classification of links. I doubt if such a classification is possible. The purpose of examples 2-4 is rather to convey a particular style of thinking.

• *Example 2*. Consider the following statements about competition.

1. Competition *is* an interaction in which individuals have negative effects upon each other by influencing access to resources.

2. Competition *is* one of the causes of evolutionary change.

Either statement provides information on competition. However, the nature of the information is very different in the two cases. In biology, the first statement would express a decision to use the word 'competition' in a particular way. It provides information on the meaning of a concept. In contrast to this, the second statement purports to deal with facts concerning competition.

Notice that the seemingly innocuous word 'is' has different meanings in the two statements. It can express a logical connection or an empirical connection. In the terminology introduced in chapter *1*, the first statement is a logical statement. Its truth does not depend on any facts uncovered by research. The second statement is empirical. To find out if it is true we will have to do research.

The example illustrates that it is not always wrong to use ambiguous words. It would be foolish to avoid the verb 'to be' because it is ambiguous. The point is rather that the intended meaning should be clear in the context. •

• *Example 3*. Biologists have occasionally quarreled about the thesis that species cannot interbreed. To understand the dispute one will have to know about the concept of species. The concept is tricky. In fact many species concepts exist. Biologists are continually proposing new ones. A famous concept for sexual organisms is the so-called biological species concept of Mayr (1970), which is introduced by him as follows. "Species are groups of interbreeding natural populations that are reproductively isolated from other such groups." This species concept implies that species do not interbreed, if the expression 'reproductively isolated' is taken to stand for *total* isolation. The thesis therefore is a *logical* statement if we adopt Mayr's concept, and interpret his definition in a rigorous fashion.

In Europe there are two species of crow, the black crow and the hooded crow, which have different colors. The black crow is all black, the hooded crow is partly black and partly grey. It turns out that intermediate forms due to interbreeding occur in various places. Would that indicate that interbreeding among species is possible after all? If we use Mayr's species concept in a rigorous way the answer is no. The reaction should rather be that it is wrong to put black crows and hooded crows in different species. Biologists have indeed decided a couple of years ago to regard them as subspecies.

Most biologists would presumably opt for a slightly different discourse concerning species—which may lead to a different decision concerning black crows and hooded crows. Mayr's species concept need not be used in a rigorous way, since reproductive isolation is a matter of degree. Species may remain distinct even if they interbreed occasionally, or more frequently but in a highly localized hybrid zone. Mayr's definition, interpreted non-rigorously, allows for limited degrees of interbreeding among species; the facts show that limited interbreeding does occur in nature. Under this interpretation the thesis that species cannot interbreed would become a false *empirical* statement.

In the evaluation of statements we obviously need to account for the context of discourse. It is important to distinguish logical and empirical statements within a particular discourse, but the distinction need not apply across different discourses. •

• *Example 4*. Consider the following quotation from Engelhardt (1981).

> ... evaluation enters into the enterprise of medical explanation because accounts of disease are immediately focussed on controling and eliminating circumstances judged to be a disvalue. The judgements are in no sense pragmatically neutral. Choosing to call a set of phenomena a disease involves a commitment to medical intervention, the assignment of the sick role, and the enlistment of health professionals.

Engelhardt is an important representative of *normativism*, a school in the philosophy of medicine which holds that medicine, at least clinical medicine, is an inherently normative discipline. This thesis would have substantial consequences for medicine. It implies, for example, that medicine cannot be a science if one assumes that science must be value-neutral.

Would it be impossible indeed to talk about health and disease in neutral terms? Before answering this question we should notice that the words 'accounts' and 'involves' in the quotation are ambiguous. An account of a disease can amount to a definition, that is a logical statement, but it can also take the form of empirical or of normative statements. Likewise, the commitment Engelhardt mentions can be 'involved' in a logical, an empirical or a normative way.

I would not accept the thesis that clinical medicine is a normative discipline if one means by this that its theories are necessarily normative. We should notice in this connection that statements such as 'Diseases *are* undesirable' are ambiguous. 'Are' in the statement represents a logical connection if being undesirable is regarded as part of the meaning of 'disease'. Alternatively, one could characterize the meaning of 'disease' in 'neutral' terms which belong to, say, biology. In that case 'are' would not represent a logical connection.

On the latter interpretation, the thesis that diseases are undesirable by no means implies that theories about disease are necessarily normative. Even if one opts for the first interpretation, the elaboration of non-normative theories concerning biological *aspects* of disease remains possible.

My comments will not suffice to settle disputes over normativism. They do show that the argument in the quotation is inadequate. •

It is convenient to have name tags for two categories of features which played a role in the examples I gave. Features which belong to the meaning of a concept are called *defining features* of the concept. Thus being an interaction among organisms (i), with negative effects (ii), via resources (iii), are defining features of a particular concept of competition. Being a cause of evolution is an *accompanying feature* of this concept, since competition is supposed to be a cause of evolution as a matter of fact (see example 2).

Biologists seldom mention this distinction. We will have to infer from the context whether particular features are meant as defining or as accompanying ones.

Defining features specify the meaning or *intension* of a term. Terms can also be characterized by their *extension* or *reference*, the class of things to which they apply. For example, fish, amphibia, reptiles, birds and mammals jointly constitute the extension of 'vertebrate'. The presence of a vertebral column is commonly regarded as a feature which belongs to the intension of 'vertebrate'.

The next example illustrates the distinction of defining and accompanying features.

• *Example 5.* Many biological phenomena (respiration of organisms, movements of plant leaves, birdsong, etc.) show rhythms with a period of 24 hours, which are associated with rhythms in environmental factors such as light and temperature. Under artificial conditions without periodicities in the environment, the period of these rhythms—barring coincidence—deviates from 24 hours. Rhythms with this feature are called circadian rhythms. The synchronization of a circadian rhythm by a periodicity in the environment is called entrainment. Deviation from a 24-hour period under constant conditions is a defining feature which belongs to the intension of 'circadian rhythm'. The possibility of synchronization with external 24-hour periods is an accompanying feature.

In biology, the thesis that all circadian rhythms have a free running period deviating from 24 hours functions as a logical statement. It makes no sense to investigate it by research. The thesis is simply true by virtue of the meaning of words. The thesis that a particular rhythm is circadian is empirical. You will need facts if you want to know whether it is true. The mere fact that the rhythm has a 24-hour period under normal conditions will not suffice for this. Non-circadian rhythms can also have this property. Hence you will need to check whether the rhythm also has the defining feature that there is a free running period deviating from 24 hours. •

Concepts come in different kinds. An important distinction in science is between *qualitative concepts* (for example 'abundant'), *comparative concepts* ('more abundant than') and *quantitative concepts* ('number of individuals per unit area').

Another distinction is between *observational concepts*, which stand for things which are open to observation, and non-observational concepts. Those in the latter category are characteristic—though not exclusively so—of theories, especially those which deal with unobservable entities. Hence observational concepts are often opposed to *theoretical concepts* which occur in theories.

In the last few decades philosophers of science have argued that *all* concepts are loaded with theoretical notions concerning unobservable entities. Thus there would not be any purely observational concepts. This has important consequences for the status of empirical science. Observation obviously cannot provide a rock bottom of indubitable evidence. This is one of the reasons why complete certainty in science is a myth.

The next example illustrates how concepts of biology are infused with theory.

• *Example 6*. Let's take the term 'observable' in a broad sense, such that it applies to the things we see through an electron microscope. Histologists who use the microscope will say that they 'see' things such as neurosecretory granules. Does this imply that 'neurosecretory granule' is an observational concept? That is a moot point. At the very least one should note that the concept has a heavy theoretical load. Those who don't know underlying theory will not recognize entities seen through the microscope as neurosecretory granules. True, with some training they could learn to identify entities which biologists call by that name, but they would need to assimilate underlying theory to grasp the full meaning of terminology. •

Exercises

2.1.1. Two biologists disagree about the question whether unicellular animals have a nervous system.

> *Biologist A*: "Unicellular animals have a system which has the function of transmitting electrical signals. This system obviously is a nervous system since it has the essential features of such a system."
> *Biologist B*: "Nervous systems are composed of cells. Unicellular animals do not have cells, so they can't have a nervous system."

Suppose the biologists ask for your opinion. What would your reaction be?

2.1.2. The following two statements have a different status from a methodological point of view. What is the nature of the difference?

> *Statement A* Predation causes an increase of diversity in prey species.
> *Statement B*. Predation causes mortality in prey species.

2.1.3. Lovelock (1979, 1988) has proposed that the earth must be regarded as an organism. He calls it by the name of Gaia. Lovelock's argument is that the earth, like organisms, has the feature of self-regulation. That is, there are mechanisms which keep conditions on the earth between narrowly defined bounds. Lovelock's theory is controversial. Many debates have been devoted to it in recent years. Would you regard Gaia as an organism?

2.1.4. The following passage from Dijkgraaf and Vonk (1971, vol. I, p. 64, translation mine) contains both logical and empirical statements. Which statements are logical, which ones are empirical?

> Vitamins are organic compounds which the animal cannot synthesize and which are present in amounts so small that they cannot play a role as material for combustion. ... Vitamins are a chemically heterogeneous group of compounds. Some of them are water-soluble, others are fat-soluble. ... Absence of vitamins results in avitaminoses, which are a form of deficiency diseases.

2.1.5. In the following passage, Goossens (1980, p. 107) criticizes strong versions of normativism in medicine. At the end of the passage, he defends a weaker form of normativism. Would his argument suffice as a defence of this view?

> Together, the nature of medicine and the nature of language make normativism extremely attractive, though not necessarily right, as a program for the analysis of medicine. Why, then, do particular normativist theses fare so poorly? One reason is that the right normative concepts have not been deployed, but the main reason is that there has been a too limited conception of how medical concepts could connect to normative ones. Some diseases, like arthritis, harm by their very presence. Others, like arteriosclerosis and the inability to make some kinds of antibodies, may produce no actual harm, but raise the probability of being harmed. Yet others, like tumors, may be dangerous for what they can lead to. What connects these medically significant conditions to well-being, however, is that they are *some threat* to well-being. ... Being a threat to well-being is proposed broadly as a necessary condition for negative medical concepts [not merely disease], but not as a sufficient condition for any particular one.

2.2. Definitions

The meaning of words can be fixed by *definitions*, that is by appeal to other words with known meanings. Here is an example, one which is typically found in philosophy rather than biology books:

'Man' =df 'featherless biped animal'.

The left hand side of the definition is the word to be clarified, the *definiendum.* The right hand side is an expression which provides the clarification, the *definiens.* The symbol '=df' indicates that this is a definition. The definiens, in this case, specifies the intension of the word defined. That is, we are dealing with an *intensional def-*

inition. Definitions which specify extensions are called *extensional definitions*. An example is

'Vertebrates' =df 'fishes, amphibians, reptiles, birds, mammals'.

Science mostly aims at intensional definitions, which are a better basis for theories.

Definitions can have two different functions, (i) stipulation and (ii) clarification or explication. Stipulative definitions simply specify the meaning of new concepts one wants to introduce. They are relatively unexciting philosophically since there often is no point in quarreling about a stipulative definition. Everybody is free to use new concepts in any way they please. Definitions meant to clarify existing concepts, *lexical definitions*, are much more interesting. We can quarrel about them since it is not easy to capture the gist of unclear language in clear terms.

Lexical definitions must accord with certain facts concerning the way words are used. Thus a definition of 'bird' may express facts concerning the actual use of the word 'bird' in biology. It does *not* express facts concerning birds. Therefore lexical definitions, like stipulative ones, are logical statements (see chapter *1*).

As the following example shows, definitions are mostly presented in informal ways in science. *Scientist seldom say whether their statements are meant as definitions or other logical statements, or as empirical statements.* Also, confusingly, some of them use the term 'definition' for descriptions which are clearly empirical.

• *Example 1*. The following passage is from Futuyma (1986, p. 7).

> BIOLOGICAL EVOLUTION ... is change in the properties of populations of organisms that transcends the lifetime of a single animal. The ontogeny of an individual is not considered evolution; individual organisms do not evolve. The changes in populations that are considered evolutionary are those that are inheritable via the genetic material from one generation to the next. Biological evolution may be slight or substantial; it embraces everything from slight changes in the proportion of different alleles within a population ... to the successive alterations that led from the earliest protoorganisms to snails, bees, giraffes, and dandelions.

In this passage Futuyma presents a lexical, intensional definition of 'biological evolution'. The first sentence clearly belongs to the definition. The second sentence merely explains what is said in the first one. The third sentence is again part of the definition; it mentions additional defining features. The fourth sentence expresses an empirical statement about biological evolution. •

Definitions are adequate only if they satisfy *definition rules*. Some important rules are listed below.

1. The meaning of the definiens should be clear.
2. Circles in definitions are not allowed. That is, the definiens should not contain the definiendum or words which are themselves defined with the definiendum.
3. Definitions should not be too broad or too narrow. That is, the definiens should not apply to more or to fewer things than the concept defined.
4. Definitions should not mention accompanying features.
5. Definitions which are unnecessarily negative are not allowed. That is, the definiens should refer to features that need to be present rather than features that need to be absent.

Rule 1 will not need clarification.

Rule 2 is clear-cut but tricky. Consider the following phrasing of the 'philosophical' definition at the beginning of this section. 'Individual x is a man' =df 'individual x is a featherless biped animal' (x is a variable). In this version of the definition, which appears to be as proper as the original one, we seem to be saddled with a circle. The original definition, however, did not contain a circle. How should we deal with this situation? The answer turns out to be simple. The expression 'individual x', the source of the trouble, does not need to be clarified; its is assumed to be clear. The matter is obvious in this rather artificial case. In realistic cases presented below and in various other places in this book, the point is not that obvious.

Rule 3 applies only to lexical definitions. Its rationale is that, in attempts to clarify concepts, *we should be faithful to the discourse considered.* If you want to clarify, say, a particular concept of altruistic behaviour *in biology*, you should not regard honorable intentions as a defining feature. Biologists do not have such a feature in mind when they are studying altruism. For them altruism has to do with effects of behavior, not intentions underlying it. A definition which mentions honorable intentions would be too broad *and* too narrow. It would apply to behaviors which are not altruistic in the biologist's sense *and* it would exclude altruistic behavior in organisms that cannot have such intentions. Notice that faithfulness has its limitations. You won't be able to clarify any concepts unless you change meanings a bit.

Rule 4 applies only to intensional definitions. Its rationale is that *we should aim to distinguish logical and empirical matters.*

Let me indulge in some fiction which explains this. Research in biology appears to indicate that all organisms do have honorable intentions after all when they show altruistic behaviour. On this assumption inclusion of 'honorable intentions' in the definition would amount to a violation of rule 4 (not necessarily of rule 3). Such intentions would constitute (allegedly so) an accompanying feature of altruism in the context of biology. If they would be taken as defining for altruism, a logical statement would take the place of an interesting empirical thesis: all organisms have honorable intentions when they show altruistic behavior! If the intentions would be defining for altruism, research aiming to check the thesis would be superfluous (see comments on logical and empirical statements in chapters *1* and *2.1*). The sequel of the story might be that biologists become convinced that the thesis is true and that they do regard further research as superfluous. As they become convinced of this, they may start using 'altruism' in a different way which does make honorable intentions defining. Thus we would be dealing with a new concept of altruism and my comments concerning rule 4 would cease to be applicable.

Rule 5 will not cause problems I guess.

I have appended miscellaneous examples to this section. Examples 2 and 3 further illustrate the application of definition rules. Examples 4 and 5 put the materials in this section in a broader context and they introduce some complications. The emphasis is on two themes. *(i) Clarity of concepts is an ideal that cannot always be realized. (ii) Clarity may be context-dependent.* These themes were already introduced in *2.1*. By now you will have noticed that *context-dependence is an important theme in this book.* It will often recur in the rest of the text.

• *Example 2.* Let's first look at some definitions which are glaringly inadequate, to illustrate the use of definition rules.
a. 'DNA' =df 'substance with an important role in the metabolism of organisms'. Too broad; meaning of definiens not sufficiently clear (cf. 'important').
b. 'House sparrow' =df 'bird species common in cities; different from starling'. Too broad *and* too narrow; mentions accompanying features; unnecessarily negative.
c. 'Organism' =df 'system which comes into existence through the reproduction of organisms'. Circle in definition. •

• *Example 3.* Rapoport and Rapoport (1960, p. 6 and p. 10) define 'biogenerative system' and 'living matter' as follows.
"... a system of factors, sufficient for the formation of living matter, including the process itself, will be called a biogenerative system ... ."
"We may offer the following definition: living matter is any system which, having been originated in a biogenerative system, is capable of maintaining a biogenerative system."
These definitions are inadequate because they are circular. •

• *Example 4.* In the process of biological classification, organisms are put into *taxa* (species, genera, families, etc.). It has been recognized since long that precise definitions of taxa names are impossible. Organisms show much variability. Therefore we can at best characterize a taxon by 'defining features' which many (not all) of its members have. Members of a taxon will have many (not all) of these features. Concepts which can only be defined with such 'defining features' are called *polythetic concepts* or *cluster concepts* (see Hull, 1965; Beckner, 1968).

Nowadays many biologists and philosophers of biology hold that taxa names must not be defined at all (see e.g. Hull, 1978; Ghiselin, 1987; Mayr, 1987). They regard taxa (at least species) as individuals which have proper names, that is names we give to particular persons and other individual entities. Such names remain undefined. •

• *Example 5*. Consider the following definition. 'Area of distribution of species x' $=_{df}$ 'set of places where members of species x occur'. There are problems with this definition but it is *not* problematic because the expression 'species x' occurs on both sides. The definiendum is 'area of distribution' rather than the whole expression at the left hand side; the meaning of 'species' is presupposed by the definition. So there is no circle.

An ecologist who investigates competition between two species A and B will do an analysis of distribution at a 'fine-grained' scale of space. If it is shown that A and B occur in the same areas, but have different habitats within the areas, the ecologist will say that they have different areas of distribution. A biogeographer who is interested in different questions may do a 'coarse-grained' analysis which supports the thesis that A and B have the same area of distribution. The expression 'places where ... occur' is obviously vague. It needs to be made more specific, but the way we do that will depend on the context, that is on the purposes we have and the questions we ask.

The concept of 'area of distribution' is useful despite its vagueness, provided we give it a more specific meaning for the context of interest. •

Exercises

2.2.1. Criticize the following definitions.

  a. 'Mammal' $=_{df}$ 'animal without feathers'.
  b. 'Cells' $=_{df}$ 'units which organisms are composed of'.
  c. 'Reproduction' $=_{df}$ 'production of descendants through the fusion of a male and a female cell'.

2.2.2. Consider the following definitions.

  a. 'Physiological process' $=_{df}$ 'chemical or physical process within an organism'.
  b. 'Organism' $=_{df}$ 'system characterized by physiological processes'.

Would you allow the concepts so defined to play a role in one theory?

2.2.3. The toxicity of chemical compounds is often determined by a cruel test in which the LD_{50} is determined. 'LD' stands for lethal dosage; the LD_{50} is the dosage at which 50% of the animals studied die as a result of exposure. Consider the following definition of 'LD$_{50}$'.

'LD$_{50}$ of compound x for species y' =$_{df}$ 'dosage of x which leads to 50% mortality in y'.

a. Is the definition circular?
b. Is the definition inadequate in other respects?
c. Replace the definition by a better one if your answer to either *a* or *b* is 'yes'.

2.2.4. Consider the following passage from a hypothetical biological article.

> The concept of fungicide is often defined as follows. A chemical compound is a fungicide if it induces mortality in fungi exposed to it. In research on fungicides, we have treated the fungus *Penicillium notatum* with ten brands of fungicide. [A description of the experimental set-up is subsequently given.] Untreated controls were included in the experiment. Four out of ten chemicals proved to have no effect at any concentration; the results for treated and untreated cultures were similar in these cases.

a. Do the experimental results indicate that the definition of 'fungicide' is wrong?
b. Do the results warrant the conclusion that some of the chemicals studied are not fungicides?

2.2.5. Suppose a biologist defines evolution as genetic change of populations. Let us accept this definition. Characterize in general terms the nature of evidence that would lead to the conclusion that evolution without genetic change is impossible.

2.3. Empirical Reference, Operationality and Coherence

So far I have only dealt with one methodological criterion which concepts must satisfy, the criterion of clarity. It is obvious that there are other criteria to be satisfied. First and foremost, concepts are useful only if we are able to assume on good grounds that they apply to the things we investigate. The criterion of *applicability* demands that it is possible to obtain information about this. The criterion can be subdivided into a criterion of *empirical reference* and a criterion of *operationality*, which I will discuss in this order.

A concept has *empirical reference* if there are concrete things to which it applies. Thus 'mammal', unlike 'unicorn', has empirical reference. The demand that concepts of science should *always* have empirical reference would be too strong. Instead one should opt for a weaker criterion of empirical reference: *concepts must have empirical reference if the context requires it.* That is a bit vague, but in actual practice it will do.

In taxonomy as practiced by biologists, 'unicorn', unlike 'mammal' will not be an acceptable concept, because it has no empirical reference. After all taxonomists want to name organisms, not fictitious entities. By contrast, a biologist who argues that for theoretical reasons no organism can be immortal, obviously needs a concept that allegedly has no empirical reference ('immortal') to express his opinion.

Two examples will further illustrate the role of empirical reference in biology.

• *Example 1.* Mayr's biological species concept (see *2.1*, example 3) has been criticized on various grounds. This concept views species as entities separated by interbreeding barriers. The most fundamental objection to this concept has been voiced by Ehrlich and Raven (1969). Their research indicated that neat populations or groups of populations separated by interbreeding barriers may simply not exist, at least not among the butterflies they studied. From a logical point of view, it is indeed possible that there are no biological species. If this were true, Mayr's concept would be inappropriate for lack of empirical reference. Most biologists, though, grant that barriers to interbreeding exist in nature. •

• *Example 2.* In evolutionary biology, 'altruism' is sometimes defined as follows. 'Behavior x of organism y is altruistic with respect to organism z' $=_{df}$ 'x decreases the fitness of y and increases the fitness of z'. In the past, the thesis has been defended that altruism in this sense cannot exist, since natural selection would work against it. Nowadays we know that this thesis is too simplistic. Suppose, however, that it were true. In that case the concept of altruism, as defined, would have no empirical reference. That would not imply that it is useless. The thesis that altruism does not exist would be very interesting. To express it we need a concept without empirical reference. •

Knowing that a concept has empirical reference is not enough. We will also want to know what the empirical reference consists in. Criteria to determine this are called *meaning criteria.* Defining features will often serve as meaning criteria. The observation that a thing has the defining features associated with a concept shows that the concept applies to it. In many cases, however, defining features do not easily permit observation, so that we will have to infer their presence or absence in an indirect way. Sometimes that can be done via the observation of accompanying features which are associated with defining ones in the relevant context. Accompanying features should thus be permitted to double as meaning criteria.

In the 1920s and 1930s, some philosophers of science have defended a thesis known as *operationalism,* which says that concepts must be operational in the sense that their defining features must be accessible to observation or direct experimental assessment. Nowadays no philosopher would defend operationalism. Even in physics, allegedly the most rigorous science, operationalism would have awkward consequences. For example, it would have the strange implication that we need to distinguish various concepts of length because lengths are measured in different ways (cf. astronomy, electron microscopy). Unfortunately, operationalism tends to linger on in biology. Hull's warnings against it (Hull, 1968) were never assimilated by the biological community.

It is true that concepts must be operational in the weaker sense that we must be able to decide on reasonable grounds whether or not they apply to particular things. This is expressed by the following criterion of operationality. *Concepts must be operational in the sense that there are meaning criteria for decisions concerning applicability in concrete cases.*

Examples 3 and 4 put the subject of operationalism in a more concrete setting.

• *Example 3*. The concept 'identical twin' is defined nowadays as 'twin originating from the splitting of a single zygote'. We will not normally be able to observe the presence or absence of the defining feature mentioned in the definition. Yet it is not difficult to distinguish identical twins from nonidentical ones, because the defining feature is strongly associated with accompanying features such as striking morphological resemblance. •

• *Example 4*. The biological species concept has often been criticized on the ground that interbreeding, its defining feature, is hard to assess by direct observation (see e.g. Sokal and Crovello, 1970). This criticism is based on an indefensible form of operationalism. The use of morphological and other features for inferences concerning interbreeding is entirely appropriate even though one can't be sure in all cases. As a matter of fact, reproductive barriers are mostly accompanied with differences in morphology, physiology and behavior. •

One criterion that concepts must satisfy remains to be mentioned, the criterion of *coherence*. According to this criterion concepts with an isolated position won't do. To be fruitful, a concept should belong to a network. In practice we can ignore this criterion since scientists will not work with isolated concepts. However, a stronger variant of the criterion, which I will call the criterion of *theoretical significance*, does deserve scrutiny. According to this criterion *we should choose our concepts in such a way that the coherence of theories is maximized.*

• *Example 5*. A natural requirement for a species concept is that it must fit biological theory. Now the theory of evolution is a central theory in biology. This calls for the elaboration of an evolutionary species concept, that is, a concept with a historical dimension that allows us to determine if organisms at different time horizons belong to the same species. Such a concept would have much theoretical significance. For example, some biologists have defined 'evolutionary species' as 'genealogical lineage between two speciation events'.

Notice that the biological species concept does not fit the bill because we can't compare non-contemporaneous organisms with respect to interbreeding. However, one could replace the notion of interbreeding by a concept of potential interbreeding which applies to organisms or populations that would have been able to interbreed if they would have been contemporaneous. A morphological yardstick applied to fossils could be used for inferences concerning potential interbreeding.

Unfortunately this expanded biological species concept is useless because in all probability it has no empirical reference. A species concept with a historical dimension would inevitably saddle us with parent species and daughter species that are adjacent in time. It is reasonable to assume that relationships of potential interbreeding would exist among such species. This would imply that no species could be an expanded biological species.

Various researchers in biology and in the philosophy of biology are now trying to elaborate a better evolutionary species concept. They appear to be motivated by the search for theoretical significance. As yet none of the concepts proposed has been generally accepted. •

Exercises

2.3.1. The concept of gene can be defined in various ways. Genes could be regarded as the smallest units of hereditary material that play a role in the transmission of features of organisms, or the smallest units that can be changed by mutation, or the smallest units that are subject to genetic recombination (the reshuffling of hereditary material on chromosomes). We could also combine these ideas, and define 'gene' as 'unit of transmission, mutation and recombination'. Biologists would not accept such a definition. Can you guess the reason for this?

2.3.2. It is difficult to measure the energy consumption of organisms in a direct way. The measurement of oxygen consumption, which mostly correlates well with energy consumption, is more easy. Would that be a reason to define 'energy consumption' as 'oxygen consumption'?

2.4. Concepts and Classifications

Concepts have intimate links with classifications. The statement that something is green presupposes that the world contains green and non-green things. Thus the concept of green is associated with a *dichotomous classification*, a classification consisting of two classes, green things and non-green things. Dichotomous classifications play an important role in our thinking which I will evaluate later on. Of course classifications often contain more than two classes (cf. red, orange, green, and so forth).

The way we should classify things will depend on the purposes we have. Many purposes are served well only if classifications satisfy the following principles.

1. *Classifications must be exclusive.* That is, no thing is allowed to belong to more than one class.
2. *Classifications must be exhaustive.* That is, each of the things in the domain of a classification must belong to some class.
3. *Criteria of classification must apply to each class in the same way.*

The role of these principles is explained by the examples below.

• *Example 1.* The classification of consumers into carnivores, herbivores and omnivores satisfies all the principles if concepts are defined in the proper way. However, if 'carnivore' were defined as 'animal which has meat in its diet', the classification would not be exclusive. A classification with only carnivores and herbivores would not be exhaustive. The classes of carnivores, herbivores and birds would form an inappropriate classification for lack of a consistent usage of classification criteria. The kind of food eaten is the criterion for distinguishing carnivores and herbivores. Birds do not fit in with this. •

• *Example 2.* In *2.3*, example 2, 'altruism' was defined as follows. 'Behavior x of organism y is altruistic with respect to organism z' $=_{df}$ 'x decreases the fitness of y and increases the fitness of z'. A natural way to define 'egoism' is by interchanging 'decreases' and 'increases' in the definiens of 'altruism'. Notice that the categories of altruistic and egoistic behaviors, if defined in this way, do not constitute an exhaustive classification. Therefore the fact that some behavior is not altruistic would not imply that it is egoistic. •

Exercises

2.4.1. Criticize the following passage from an article by Hairston, Smith and Slobodkin (1960).

> ... cases of obvious depletion of green plants by herbivores are exceptions of the general picture, in which the plants are abundant and largely intact. Moreover cases of obvious mass destruction by metereological catastrophes are exceptional in most areas. Taken together, these observations mean that producers are neither herbivore-limited nor catastrophe-limited, and must therefore be limited by their own exhaustion of a resource.

2.4.2. Eldredge and Gould (1972), in a famous article, distinguish two modes of speciation, which they define as follows.

Phyletic gradualism
1. New species arise by the transformation of an ancestral population into its modified descendants.
2. The transformation is even and slow.
3. The transformation involves large numbers, usually the entire ancestral population.
4. The transformation occurs over all or a large part of the ancestral species' geographic range. (....)

Punctuated equilibria
1. New species arise by the splitting of lineages.
2. New species develop rapidly.
3. A small sub-population of the ancestral form gives rise to the new species.
4. The new species originates in a very small part of the ancestral species' geographical extent—in an isolated area at the periphery of the range.

Suppose evidence warrants the rejection of phyletic gradualism. Would that imply that punctuated equilibria, the view cherished by Eldredge and Gould, is right?

2.5. Hard Cases

Concepts which describe mental events, or events with mental aspects, are notoriously hard to define even though we use them without problems in *everyday discourse*. We all know what 'seeing' amounts to, or so it seems, but it is difficult to define the verb 'to see' in a precise way.

Would biology offer a way out? It is tempting to define 'seeing' as a process involving the formation of patterns on the retina, which lead to 'images' in the brain. However, there are two problems with this. First, this definition would fail to capture the subjectively experienced, mental aspect of seeing. Secondly, one wonders which entity is 'seeing' the image in the brain.

If we feel that the concept of seeing is nonetheless clear to us, we need not despair if we are faced with these problems. After all, *it is impossible to provide definitions for all the concepts we use*. An attempt to do this will ultimately lead to circles in definitions, so some concepts will have to be introduced without defining them, that is as *primitive concepts*. Perhaps the concept of seeing is best regarded as a primitive one in most contexts.

There is another problem with the concept which will only come to the surface by a *scientific approach*. As explained in the example below, seeing appears to be an amalgam of different processes which we experience as one process in *daily life*. This should not count against the normal usage of the concept of seeing, but it does indicate that an intuitive interpretation of subjective experience has its limitations.

• *Example 1*. Patients with a certain form of brain damage have an impaired visual field, that is there are blind parts in it. In Weiskrantz et al. (1974) a surprising experiment with one patient is described. This patient was able to 'guess' correctly what objects were placed in the blind part of his visual field, but he would deny that he saw them (for a detailed discussion, see Humphries and Riddoch, 1987, and Weiskrantz, 1987). The guesses of the patient could be described as 'seeing without knowing that one sees'. Seeing is apparently a combination of at least two processes, a conscious and an unconscious one. In normal seeing these processes are tightly coupled. In pathological conditions they may become dissociated. Ordinary language does not suffice to describe what happens under such conditions. •

Concepts representing properties or relations can have various degrees of complexity. The simplest concepts are expressed by so-called *one-place predicates*. An example is 'hot', which can be used in expressions with the form 'x is hot'. The concept is tied to one item (x), hence it is expressed by a 'one-place-predicate'. 'Hotter than' is more complex, it is a *two-place-predicate* (cf. 'x is hotter than y'). 'Gives' is a three-place-predicate (cf. 'x gives y to z'). Many concepts of biology are complex, that is they should be expressed with *many-place predicates*, predicates with more than one place. Complex concepts are often not recognized for what they are. Various examples presented below show that this can have unfortunate consequences (for additional examples, see Van der Steen, 1990a).

• *Example 2*. Various features of organisms are said to be genetically determined. An example is eye color in man. Casual inspection of the concept of genetic determination suggests that it is harmless and clear. However, the following line of reasoning shows that there are pitfalls. On the one hand, intuition indicates that eye color differs from certain other features in being genetically determined. On the other hand, biology teaches us that *all* features of organisms are determined by genetic *and* environmental factors. Accordingly, the statement that eye color is genetically determined should not make sense because it suggests that some features are not so determined.

The solution of this puzzle is that we should not construe 'genetic determination' as a one-place predicate but as a three-place predicate. A *difference* in a feature between two organisms can be genetically determined in the sense that it results from a genetic difference between them (for details, see Voorzanger, 1987, and Gifford, 1990). Hence we get the following definition. 'The difference in x between y and z is genetically determined' $=_{df}$ 'the difference in x between y and z is caused by a genetic difference between y and z'.

Therefore the genetic determination of a feature will depend on the comparison one makes. Skin color is genetically determined under some comparisons (cf. differences between people belonging to different races), and environmentally determined under different comparisons (cf. people who stayed at home in Canada and people who went to the Bahamas).

Our intuition about eye color can be understood as follows. Eye color (in adults) shows genetic determination in a strong sense of the term: differences in eye color between people are the result of genetic differences under *all* comparisons. Few features are genetically determined in this strong sense.

The informal usage of terms such as 'genetic determination' and 'environmental determination' can have serious consequences. For example, some researchers who found a genetic difference between (some) patients with a psychiatric disorder and (some) healthy persons, have concluded that the science of psychiatry should be founded on biology since 'the' causes of disorders are genetic. There is no warrant for this view, which is invoked to support a one-sided pharmacological treatment of patients (for further comments, see Van der Steen and Thung, 1988, and Van der Steen, 1991, 1993b). •

'Genetic determination *versus* environmental determination' is an example of a misleading, even harmful dichotomous classification. *False dichotomies are common in the literature. They represent a tendency of human beings to resort to black-or-white thinking.* Another common phenomenon is that different dichotomies are amalgamated. As the following example shows, this can lead to *unacceptable forms of bias.*

• *Example 3.* Schwartz and Africa (1984) offer the following classification of etiological factors (= causes of disease) of schizophrenia.

A Biological factors
    A1 Genetic factors
    A2 Specific abnormalities
        A2a Anatomical and physiological factors
        A2b Biochemical factors
B Psychosocial factors
    B1 Development of the individual
    B2 Development within the family
    B3 Development within society and the larger environment
        B3a Population density
        B3b Socioeconomic class
        B3c Date of birth
        B3d Other factors

Notice that, for Schwartz and Africa, biology is concerned with the study of processes inside the organism. The environment is represented by psychosocial factors. The authors apparently merge two dichotomies, biological *versus* psychosocial factors and internal *versus* external factors. The ensuing classification is not exhaustive (see *2.4*) since environmental factors in the province of ecology are disregarded (population density being an exception). Now it has been shown by some researchers that factors such as the composition of diet can be implicated in the genesis of psychiatric disorders. Many standard texts in psychiatry do not mention this possibility. Classifications like the one given by Schwartz and Africa are common in medicine (for additional examples, see Van der Steen and Thung, 1988, and Van der Steen, 1991, 1993b). We can but wonder about the consequences for medical treatments. •

Philosophers have recognized since long that *concepts of science* are complex because they *are loaded with theory*. The concepts analysed above are an example. In many cases we will have to accept complexity, and try to be clear about it. In other cases, however, scientists appear to use concepts I would like to de-complexify. The next example clarifies what I mean by this.

• *Example 4*. Several decades ago, the researcher Selye (for reference see Selye, 1983) discovered in physiological studies that many different, harmful environmental stimuli can trigger a response he called the *stress response*. The response consists in various physiological changes, for example in heart rate, skin resistance and hormone concentrations. Under prolonged exposure the response can wane by an adaptation process, but that does not mean that all is well. In the long run disease and even death may ensue. In his work Selye concentrated on responses. Psychologists investigated stress in the same period. Their emphasis was on stimuli and internal states (for example emotions) associated with the stress response.

Later on researchers in psychosomatic medicine tried to integrate biology and psychology in research on stress. It was felt that an integrative stress *concept* is needed. Accordingly 'stress' was conceptualized as a particular relation between stimuli, internal states and responses. At the same time the relation was investigated in experiments. This resulted in an enormous confusion because the relation was regarded as logical and empirical at the same time. I will simplify a bit, and forget about internal states, to illustrate problems facing researchers (for a more detailed analysis, see Van der Steen and Thung, 1988).

Suppose we define 'stress response' like Selye did, as a response characterized by certain physiological changes. Now 'stress stimulus' or 'stressor' can be defined as any stimulus that happens to produce a stress response. Possible examples are extreme heat, bacterial infection and death of a spouse. *If* we adopt these definitions, we cannot at the same time investigate the general thesis that stressors produce a stress response in empirical work. The definitions simply will not permit the discovery of exceptions to the relation, so empirical studies into this relation cannot yield anything new. One can investigate, though, which stimuli are stressors as a matter of fact.

A fundamental, empirical relation between stressors as a category and the stress response could be elaborated only if 'stressor' were given an intensional definition which does not depend on 'stress response'. However, it turns out that this is not a feasible option because 'stressor' represents a heterogeneous category.

In view of all this, 'stressor' is best regarded as a convenient label for this category. The concept cannot play an important theoretical role. If we want to chart empirical relations between features of stimuli and features of responses, we should resort to separate concepts for different stimuli, and formulate diverse, specific regularities. There cannot be a general law which connects stimuli and responses.

In short, to describe stimulus-response regularities, we should not use 'stressor' as a general, overarching predicate which unites stimuli and responses. Instead we need more specific predicates, without 'stress response' in the definition, for separate stimuli. The concept of 'stressor' is loaded with too much theory. For some purposes it is wise to replace it by specific concepts with less theoretical baggage. •

Exercises

2.5.1. As shown in example 2, the concept of genetic determination used in biology is complex. Of course we are free to adopt a simpler concept provided that we do not use it to explicate an existing concept. Suppose I opt for the following definition. 'Feature *x* is genetically determined' =df '*x* is caused by genetic factors only, that is no environmental factors have a causal role in the origin of *x*'. Could my concept of genetic determination play a role in biological theory?

2.5.2. Criticize the following passage from Mikhail (1985).

> There is no logical incompatibility that stands in the way of integrating psychological stress theory with Selye's theory. Both approaches are, in fact, complementary. Psychological stress theory outlines the conditions which determine the evocation of stress while Selye's theory describes its form. To portray what is significant in these approaches, the following definition of stress is suggested. ... Stress is a state which arises from an actual or perceived demand-capability imbalance in the organism's vital adjustment actions and which is partially manifested by a nonspecific response. An objective of this new definition is to emphasize the continuity between psychological and physiological theorizing (p. 37).

2.5.3. Criticize the following definitions, which were taken from a seminal article by Gould and Lewontin (1979).

> 'Adaptation'—the good fit [fit relating to 'design'] of organisms to their environment—can occur at three hierarchical levels with different causes. It is unfortunate that our language has focused on the common result and called all three phenomena 'adaptation' ... . First, we have what physiologists call 'adaptation': the phenotypic plasticity that permits organisms to mould their form to prevailing circumstances during ontogeny. ... Secondly, we have a 'heritable' form of non-Darwinian adaptation in humans ...: cultural adaptation ... . Finally, we have adaptation arising from the conventional Darwinian mechanism of selection upon genetic variation (pp. 592-593).

2.6. Afterthoughts

My approach in this chapter has been straightforward and practical. I could only do that by taking sides in philosophical matters which are anything but straightforward within philosophy. Therefore some justification is called for. Various other chapters will likewise be concluded by justificatory comments.

The following interconnected themes from the philosophy of science bear on the materials in the present chapter.

1. Philosophers agree that all concepts of science are loaded with theory.
2. Since everything hangs together in science, many philosophers nowadays hold that one cannot distinguish between logical and empirical matters.
3. The classical theory of meaning and reference is heavily attacked nowadays by various philosophers. According to this theory, the meaning (intension) of a concept determines its reference (extension). That is, in order to know if a concept applies to a thing one has to know if the thing has the defining features associated with the concept. I have adopted the classical theory in this chapter.

Concerning the first issue I have emphasized a possibility which does not play the role it deserves in philosophy, *viz.* that we need not take the load of concepts for granted. We can choose to manipulate the load of a problematic concept and so replace it by a better one. General concepts concerning stress (see *2.5*, example 4) are an example. If it turns out that such concepts cannot play a fruitful role in theories of biology and medicine because they are loaded with too much theory, we can choose to replace them by more specific, simpler concepts.

This has obvious consequences for the distinction of logical and empirical matters. I would defend the thesis that this distinction can be made, to some extent at least, in some contexts. Indeed I think we *should* make it to promote clarity whenever that is possible. In publications on stress it is often unclear whether links between stimuli and responses are logical or empirical. I would argue that a confusion of logical and empirical matters is by no means necessary in this case. If scientists conflate them, we could take the view that their conceptual equipment is in need of improvement.

The third issue is the hardest one. Philosophers rejecting the classical view of meaning and reference have argued as follows. The meaning of concepts in science is often subject to changes even when it is reasonable to assume that they keep referring to the same entities. This would imply that meaning cannot determine reference.

Let me illustrate this by a concrete example. The term 'gene' has been used since long for entities which 'determine' features of organisms. That is, 'gene' has been defined by a specification of the role or function that genes have. In the past it has been thought that genes are proteins. Later on biologists came to realize that they are nucleic acids. Now some biologists originally mentioned material composition as an additional feature in the *definition* of 'gene'. Those who did this were forced to change the meaning of 'gene' when it was discovered that the role attributed to genes is played by nucleic acids. It could be argued that the reference of 'gene' stayed the same all along. Genes kept being entities, whatever their nature, which play a particular role in the organism. Meanings changed, reference did not, hence meaning does not determine reference.

My reaction to this example is as follows. Biologists can opt for different definitions of 'gene' (functional *versus* functional-material) even when they agree about facts. The same findings apparently can be conceptualized in different ways. Concerning the example it seems that the plausibility of any particular philosophical view of meaning and reference will depend on the way concepts happen to be used by scientists (or reconstructed by philosophers). The story of biologists who defined genes as functional entities is apparently compatible with the classical theory of meaning and reference. Other terminologies less easily fit this theory.This seems to indicate that in philosophy, as in science, the value of models and theories is context-dependent (see also Sloep and Van der Steen, 1988).

Concerning the example above from genetics I opted for a reconstruction which minimizes conceptual change. In other episodes of science conceptual changes may play a more substantial role. Also, scientist at any particular time may use concepts in different ways. This indicates that the distinction of facts and meanings (conceptual and logical matters) is problematic if we apply it across discourses. I grant this, but as I have argued in this chapter it is possible, indeed important, to retain the distinction in the evaluation of scientific work *within* the context of a particular discourse.

It should be obvious that philosophical theories, like scientific ones, have limitations with respect to validity. This need not make them useless. *Limitations can be accommodated by applying theories only in contexts where the consequences of limitations are comparatively harmless.* Anyhow it is obvious to me that the classical theory of meaning and reference is useful in many contexts.

CHAPTER 3

Arguments and Fallacies

3.1. Deduction and Induction

Suppose you are convinced that your Aunt Beth did not die a natural death. Specifically, you think she was murdered by a visitor who put arsenic into her coffee. The police are skeptical. How could you convince them? Let's suppose that they give you the benefit of the doubt and launch an investigation. Arsenic has a tendency to accumulate in hair roots, so some hair is obtained from the deceased and given to a chemist. An analysis performed by her shows that there are traces of arsenic in the hair. The police conclude that Aunt Beth did die from arsenical poisoning. So you have made your point concerning the poisoning. Additional evidence is needed, of course, to show it was murder.

In this example a thesis—Aunt Beth died from arsenical poisoning—is substantiated in a convincing way. Precisely what makes it convincing? A superficial answer would be that the facts are convincing. The facts bear on the thesis in a rather indirect way, so we need to know *why* they substantiate it. The crucial point is that the thesis follows from statements which appear to be true. It is the conclusion of a good argument.

Arguments consist of two parts, a *conclusion*, the statement which is in need of support, and *premises*, statements which provide the support. The present argument can be reconstructed as follows. To begin with we have the premise that if any deceased person has arsenic in hair roots, then their cause of death was arsenical poisoning. Add to this the premise that the hair roots of Aunt Beth, deceased, contain arsenic, and you can infer the conclusion that the cause of death of Aunt Beth was arsenical poisoning. (I have not bothered about details. For example, arsenic would have to be present in the hair roots in certain amounts.)

The first premise is obviously based on research done in the past. Let us assume that it is reasonable to accept it as true. The second premise in the story is true since it matches the facts uncovered. On the basis of this we are entitled to accept the conclusion because the argument exemplifies a strong relationship of support. The conclusion *follows from* the premises. That is, *it is impossible that the conclusion is false if the premises are true.*

Arguments with this property are called *valid deductive arguments*, valid arguments for short. The term 'deductive' stands for the assumption that the conclusion follows from the premises. The term 'valid' indicates that this assumption is right. A deductive argument is *invalid* in case the assumption is wrong.

The study of validity belongs to logic because *one only needs to inspect the form of an argument to decide whether it is valid*. In logic, arguments which are valid and have the additional property that the premises are true, are called *sound*. Naturally, soundness cannot be evaluated with the resources of logic alone.

Notice that the concept of validity, as used in the example, does not represent a species of truth. *Validity and truth are often confused.* In my experience it is extremely hard for students of the life sciences to distinguish them in a logically proper way. Validity concerns formal relations among statements. Truth, or falsity as the case may be, is a feature of the statements so related. Validity can be defined, though, in terms of truth and falsity: an argument is valid if it is logically impossible that its conclusion is false *if* its premises are true.

The argument considered above has the following form: for all x, if x is A, then x is B; a is A; therefore: a is B. Here 'x' is a variable, 'A' means 'a deceased person with arsenic in hair roots', 'B' means 'a person who died from arsenic poisoning', 'a' is a constant representing 'Aunt Beth'. With respect to the validity of the argument, death, arsenic and Aunt Beth are immaterial. If 'A', 'B' and 'a' are given different meanings the result will again be a valid argument.

The argument concerning Aunt Beth *was set up with the purpose of defending a particular statement. Arguments can also have a different purpose. They can help us understand why a statement we already endorse is true.*

By way of an illustration I will look at the Aunt Beth episode in a slightly different way. Let's start with the undisputed statement that she died. You want to know why she died, so you start an investigation. Suspicions arise, the police is called in, and so forth. Why did Aunt Beth die? The short answer is that she died of arsenic poisoning. This answer can be regarded as a summary of the following valid argument. If a person ingests arsenic, then this person will die; Aunt Beth ingested arsenic; therefore Aunt Beth died. (I have not bothered about details concerning the dosage needed, absence of medical intervention, and so forth.)

The point of this argument is not to show *that* Aunt Beth died. We knew that all along. The argument shows *why* she died. It provides an *explanation*. In science, arguments often play a role in explanations. They are also important in prediction, hypothesis testing, and so forth. The latter uses of arguments, and details concerning explanation, are considered in later chapters. The present chapter and the next one only deal with general aspects of argumentation.

In ordinary and in scientific discourse arguments are seldom presented in a fully explicit form. For example, there often is no need to explicitly formulate premises that represent shared knowledge. *If premises of an argument are kept implicit, we will have to reconstruct it before evaluating it.*

An example of reconstruction is presented below.

• *Example.* Suppose you want to know why there are no lichens in the area where you live. A biologist you ask about this says that there is acid rain in the area.

For convenience, I will use the following symbols to deal with this example. 'A' will stand for 'an area with acid rain', 'B' for 'an area without lichens', 'a' will name a particular area, the one you live in, 'x' will represent a variable.

The biologist's answer points to a cause, acid rain, and its effect, absence of lichens. He presumably will be able to formulate an argument with the *conclusion* that a is A. Now this conclusion does not follow from the premise 'a is B' alone. If we want to get a valid argument we need to add the premise 'for all x, if x is B, then x is A'. In words, the thesis that all areas without lichens have acid rain, together with the statement that a particular area has no lichens, inevitably leads to the conclusion that this area has acid rain. The chances are that the biologist will have taken the missing premise for granted. Notice that validity is not enough for the argument to be acceptable. In addition there must be good grounds for assuming that the premises are true.

The argument in my reconstruction moves from effect to cause. Now causes explain effects, but effects do not explain causes. If the biologist's response is taken to be an explanation it is more natural to reconstruct it as an argument that moves from cause to effect. He could also have said: 'The area has acid rain, and if there is acid rain in an area it will not have lichens'. This argument can be reconstructed as one which has two premises, 'a is A' and 'for all x, if x is A then x is B', which lead to the conclusion 'a is B'.

Notice that this argument is not meant to show *that* there are no lichens in the area. We knew that already. The point is rather that we want to know *why* there are no lichens.

The most difficult step in the second reconstruction is the identification of the conclusion. We have to realize that statements which we know are true need not be taken as premises. The reconstruction can be defended as follows in a general way. The starting point is that one asks why a particular phenomenon has occurred. This can be rephrased as the question why a statement describing the phenomenon is true. An obvious way to answer this question is by showing that the statement follows from premises which we assume are true.

In the reconstruction, a causal relationship was covered by the expression 'if ... then ...'. This expression is actually incapable of covering all the connotations of causal language; for example, it does not express temporal relationships. We have to take this into account when we evaluate the original argument with the help of the reconstruction. Reconstructions or translations in which *nothing* gets lost are impossible. •

Deductive arguments can be classified according to three criteria. The premises can be true or false, the conclusion can be true or false, and the argument as a whole can be valid or invalid. This yields 2^3 = 8 classes. Only one of these classes (premises true, conclusion false, argument valid) is empty. It is easy to elaborate examples of arguments in the seven remaining classes. Sound arguments are all in one class (premises true, argument valid; conclusion therefore true).

There are also arguments with premises which are not meant to support the conclusion in a rigorous way. Such arguments should not be regarded as deductive ones. They are more appropriately called *inductive arguments*. If we study swans, and continually come across white ones only, we can reasonably conclude that all swans are white. This would amount to an inductive argument with premises expressing the whiteness of individual swans. The conclusion of the argument does not follow with certainty from the premises. It happens to be false, by the way.

For inductive arguments I use the terms 'acceptable' and 'unacceptable' rather than 'valid' and 'invalid'. Acceptability depends on many factors which I will consider later on. (To avoid proliferation of technical terms I also use these terms for 'ordinary' notions which will be clear from the context.)

The distinction between invalid deductive arguments and inductive ones is a matter of context. If someone would present the swan-argument as a valid deductive argument, your response should be that it is actually invalid if construed as a deductive argument. Most people, however, would regard the argument as an inductive one.

Exercises

3.1.1. Someone asks a biologist why willow warblers that breed in the temperate zone are migrants. The biologist replies that the they migrate because they are insectivores. Reconstruct this response as a valid deductive argument.

3.1.2. A biologist studying freshwater snails in ponds discovers that there are no snails at all in some ponds. Chemical analysis shows that the water in these ponds has a very low calcium concentration. She concludes that freshwater snails do not occur in ponds with a low calcium concentration. Would the inference of the conclusion involve deduction or induction?

   The conclusion of this argument can be used as a premise in a different one, to show why there are no freshwater snails in a particular pond which is subsequently investigated. The latter argument will work only if one adds a premise. Which premise should be added? Is the argument deductive or is it inductive?

3.2. Evaluating Arguments

Arguments are acceptable only if they satisfy various *methodological criteria*. First of all, the concepts they contain need to satisfy various criteria discussed in the previous chapter.

The criterion of *clarity* applies to concepts and arguments alike. Clarity of arguments is achieved by reconstruction, as shown in the previous section. Two additional criteria, validity and truth of premises, were also mentioned in this section. These *criteria must be stated with care*. *Validity* only applies to deductive arguments. A weaker criterion, *inductive support*, applies to inductive arguments. In valid deductive arguments the premises (if true) make the conclusion inevitable; the premises of acceptable inductive arguments only make the conclusion probable.

The requirement that premises of arguments must be true also needs to be weakened. In many cases it is impossible to know if a statement is true. Thus universal statements, that is statements with the clause 'for all x ...', cover infinitely many instances. We could never investigate them all. Hence the criterion of *confirmation*, which is weaker than the requirement of truth: the premises of arguments must be well-confirmed.

Arguments must also satisfy the criterion of *non-circularity*. An argument with the form 'p, ...; therefore: p', which has the conclusion in the premises, clearly will not do. It is valid, but it does not represent any genuine support. So we have to add the clause that *arguments in which 'support' derives from their own conclusion are inadequate. Such arguments are called circular*. The most blatant form of this is the so-called *logical circle*, which does have the conclusion in the premises. A more subtle variant is the *epistemic circle*, which I will discuss in the next section.

Next, logical statements will not do as premises which are meant to support the conclusion of an argument. Hence the criterion of *empirical content*, which says that *supporting premises must be empirical*. Needless to say, this criterion is appropriate in the context of natural science but not in the context of, say, mathematics. As the next example shows, *the criterion is closely connected with the criterion of non-circularity*.

• *Example 1*. It is conceivable that you will come across the following argument in evolutionary biology.

> *Premise 1*. If two individuals belonging to the same species differ in fitness, the individual with the highest fitness will have better chances of survival and reproduction than the individual with the lowest fitness.
> *Premise 2*. Individuals A and B belong to the same species; A has a higher fitness than B.
> *Conclusion*. A has better chances of survival and reproduction than B.

Let us assume that the argument is sufficiently clear, and that the premises are true. (The assumption is actually problematic. One should always relate fitnesses to environments. I have avoided complications which are not important in the present context.) The argument is also valid. Yet it need not be acceptable. Some biologists—by no means all of them—would regard premise 1 as a definition of 'fitness'. Let us suppose that the concept of fitness is indeed used in this way, such that '*A* has a higher fitness than *B*' and '*A* has better chances of survival and reproduction than *B*' have identical meanings. In that case premise 1 is a logical statement which does not qualify as evidence in favor of the conclusion. The argument can also be rejected on the ground that it is a logical circle since premise 2 and the conclusion have the same meaning. They represent the same statement though they are worded in different ways. •

Notice that I have worded the criterion of empirical content in a careful way. Premises which are meant to support the conclusion of an argument must have empirical content. Logical statements may be admissible if they do not have the function of support. However, as indicated in the next example such statements are in fact redundant.

• *Example 2.* A valid argument with the following pattern may well be acceptable even if the first premise represents a definition.

> *Premise 1.*   Persons with symptoms *S* have disease *D*, and *vice versa.*
> *Premise 2.*   Persons with symptoms *S* cannot be cured.
> *Premise 3.*   John has disease *D*.
> *Conclusion.*   John cannot be cured.

Notice that we can delete the first premise without loss of validity, provided that we replace 'symptoms *S*' by 'disease *D*' in the second premise, or 'disease *D*' by 'symptoms *S*' in the third one. •

Lastly, arguments must satisfy the criterion of *coherence*, which says that their premises must cohere with available knowledge. Examples 3-5 show that we do need such a criterion.

• *Example 3.* In The Netherlands, indeed in the greater part of Europe, nightingales are only present in summer, not in winter. Suppose the following argument is presented as an explanation.

> *Premise 1.*   Nightingales are present if and only if there are many people on the beaches.
> *Premise 2.*   In summer there are many people on the beaches, in winter there are not.
> *Conclusion.*   Nightingales are present in summer, not in winter.

This is obviously an unacceptable argument. Why should that be so? Notice first that the argument is valid and non-circular, and that the premises have empirical content. Also, barring details, we can assume that the premises are true. The point is obviously that the premises are irrelevant since they do not cohere with knowledge represented by biological theory or any other kind of knowledge. Therefore we can safely assume that people visiting beaches have nothing to do with the presence or absence of nightingales. •

The inadequacy of the argument in example 3 is obvious. As the next example shows, there are also more subtle forms of lack of coherence which easily go undetected.

• *Example 4*. Mary has a particular disease. She is treated with a drug, and recovers within a week. Her recovery could be explained on the basis of two premises, (i) people with the disease who take the drug always get well and (ii) Mary took the drug. This explanation would not be acceptable if it turns out that people who don't take the drug recover as well. In that case the first premise would not cohere in a substantive way with available knowledge. Recovery, in this case, would have nothing to do with the drug. •

The example below shows that new information can make us reject the conclusion of an inductive argument we deemed acceptable before the information became available. New information affects valid deductive arguments in a different way. It will not lead to the rejection of a conclusion as long as we accept the original premises, but it can make premises irrelevant (see the foregoing example). In either case the criterion of coherence demands that we take available knowledge into account.

• *Example 5*. Suppose a patient has a rare malignant tumour which is removed by surgery. If we know that, say, 90% of the patients with this tumour will recover completely after surgery, it is reasonable to infer by induction that this particular patient will probably recover. However, the argument will not do if it is known in addition that the chances of recovery are slight if there are metastases, and that the patient does have metastases. The premises would become less relevant in this case because there is specific information which better coheres with available knowledge. •

Exercise

3.2. People who have been in close contact with others who have an infectious disease, may themselves get ill through infection. However, there are those who are not susceptible to the disease. Suppose John has been in close contact with people with an infectious disease, and that he stays healthy. A physician explains this by the assumption that John is not susceptible. Reconstruct the explanation as an argument. Would you accept the explanation?

3.3. Fallacies

Arguments which are unacceptable are called *fallacies*. Logicians have distinguished many kinds of fallacy. Their inadequacy in all cases stems from a failure to meet the criteria discussed in the previous section. Arguments which are invalid are also called *formal fallacies*. *Material fallacies* are arguments which do not satisfy other criteria. (I am adopting a common terminology, but the distinction is somewhat misleading. Some of the material fallacies which have been distinguished can also be reconstructed as formal fallacies.) Some well-known kinds of material fallacy are surveyed below.

Fallacies of ambiguity deserve special attention. As the name indicates they are unacceptable because they contain an expression which is used in different ways. As indicated in chapter 2, ambiguities may cause subtle problems in science. So it should not come as a surprise that fallacies of ambiguity may be quite difficult to detect. Some authors have rejected the theory of evolution with arguments that belong to this category without noticing the ambiguity! This is worked out in the example below.

• *Example 1.* Consider the following argument.

> The theory of evolution does not allow the derivation of predictions. Theories are
> adequate only if they allow the derivation of predictions. Therefore, the theory of
> evolution is not an adequate theory.

This argument has found its way in literature dealing with the status of evolutionary biology. Notice first that it is seemingly valid. Would the premises be acceptable? The idea of the first premise is presumably that the course taken by evolution in the future cannot be known beforehand. For the sake of argument, let us take this for granted. What about the second premise? This premise is obviously a thesis which belongs to methodology. Because it is normative the concepts of truth and falsity may not apply to it, but that does not preclude an evaluation in terms of acceptability. Now methodologists definitely would not mean by it that theories are adequate only if they can help us know the future. The idea behind the second premise is rather that theories must have predictive force in the sense that they permit the derivation of new information. For example, on the basis of evolutionary thinking one could hypothesize that organisms which are intermediate between known fossils must have existed in the past. This 'prediction' may be borne out by subsequent research.

If the argument is interpreted in this way, it is a fallacy of ambiguity since the term 'prediction' is used in two different ways. Notice that it can be reconstructed as a formal fallacy on this interpretation. One term, 'prediction', is used for two concepts. Instead one should use two different terms. If the terminology is replaced by a more adequate one, the conclusion no longer appears to follow from the premises.

The first premise could also be taken to mean that evolutionary theory does not permit the derivation of new information. On that interpretation there is no ambiguity. The argument should be rejected in that case on the ground that the first premise is plainly false. •

In an *argumentum ad hominem* the premises wrongly judge persons rather than stating the issues at hand. If you are acquainted with political meetings you may know about this fallacy. Scientists usually refrain from committing it during official parts of meetings, and in publications. But we should take heed here as well. The next example makes this fallacy more concrete.

• *Example 2.* 'It is not true that the periodicity shown by circadian rhythms under constant conditions is due to environmental rhythms, because researchers defending this thesis are unable to perform correct statistical analyses.' This argument is an *argumentum ad hominem* if it is formulated in this way. However, if one works out the idea behind it, it turns out to be sound. Let me explain.

Most people doing rhythm research hold the following view. Circadian rhythms have the defining feature that their period deviates from 24 hrs under constant conditions. Their 'free running period' may have values between 20 and 28 hrs. Under normal conditions the rhythms are synchronized by 24-hr periodicities in the environment. There is evidence that free running periods, as periods, are generated by mechanisms in the organism. In other words, circadian rhythms are regarded as endogenous.

Some dissenters have argued that organisms are influenced by subtle factors in the environment such as magnetic fields and cosmic rays (which is true). They have also maintained that such factors have periodicities which easily escape detection, and that these periodicities are responsible for free running periods. Accordingly, they regard circadian rhythms as exogenous.

It can be shown that the evidence produced by the dissenters is no good, because of statistical flaws. They do not seem to have mastered elementary statistics. Accusing them of this, of course, does not settle the issue. The fact that their statistics *are* flawed as a matter of fact indicates that they have not made their case. That in itself need not imply that their thesis is false. However, the burden of proof is theirs in view of evidence indicating that circadian rhythms are endogenous. All in all, therefore, it is reasonable to reject the thesis of the dissenters.

Defenders of the standard view of circadian rhythms have occasionally been guilty of accusations in the form of an *argumentum ad hominem*. Nothing much hinges on this in the present case. In other cases, though, issues may become clouded if sensible theses are made suspect by an attack on persons. •

In an *argumentum ad ignorantiam* it is suggested that something is true (or false, as the case may be) because it has not been shown to be false (true). As the next example shows we should be careful with the label 'it has not been shown'. It implies that there is no evidence. That alone is not sufficient to brandish an argument as a fallacy. Crucial are the efforts which have gone into the search for evidence.

• *Example 3*. The following arguments have the pattern of an *argumentum ad ignorantiam*.

All features of organisms have adaptive value, because no feature has been shown to lack adaptive value.

The mortality of fish in this river is not due to pollution by the steel plant, because there is no evidence of pollution.

If the conclusion of an argument with this pattern is based on mere lack of evidence, the argument is indeed an *argumentum ad ignorantiam*. However, if evidence has not been forthcoming despite efforts to unearth it, the argument may be regarded as an acceptable inductive one. •

In an *argumentum ad consequentiam* the premises indicate that accepting the conclusion would have positive consequences, or that rejecting it would have negative consequences. This fallacy is extremely common. Here is an example.

• *Example 4*. 'Investments in environmental protection are undesirable because they would increase prices and so stimulate social conflict.' This is an obvious example of an *argumentum ad consequentiam*. If we have to decide on a course of action, we had better evaluate positive *and* negative consequences of alternatives which are possible. •

The last fallacy I will discuss is the *petitio principii* or *circular argument*. The *logical circle*, in which the conclusion occurs among the premises, is the strongest form of this fallacy (see the previous section). It is also possible that the conclusion is the only source of evidence indicating that a premise is true or acceptable. In that case the argument is an *epistemic circle*. Here is an example concerning ethology.

• *Example 5*. In ethology, concepts such as 'drive' and 'instinct' easily lead to circles. The argument that birds build nests because they have a nest-building instinct is no good if the instinct is merely defined as a disposition to build nests, the disposition being equated with the occurrence of nest-building under appropriate circumstances. If that is done the argument is a logical circle.

Other tricky concepts in ethology are 'search image' and 'innate releasing mechanism'. Animals often show a preference for food items they earlier came across and disregard other items they could eat as well. The preference is sometimes explained as due to a search image formed by previous encounters with the same food. If there is no evidence for the search image apart from the observed preference, the explanation will amount to an epistemic circle. Likewise for innate releasing mechanisms which are supposed to explain specific responses to specific stimuli. •

The examples may suggest that the distinction between circles and acceptable arguments is quite straightforward, but that is not so.

Suppose experiments indicate that ten species of green plants do not perform photosynthesis during the night. We could infer from this by induction that *all* green plants do not photosynthesize during the night. This generalization could explain the absence of photosynthesis in other species which are subsequently studied. The explanation, which admittedly is not very impressive, does not involve circular reasoning.

Could the data concerning the ten plants that were studied first be explained in the same way? In this case data would be explained by a premise inferred from the same data. This apparently amounts to an epistemic circle. However, we could also reason as follows. Part of the data could suffice to infer our generalization. So, if we explain the remaining data by the generalization, we are not guilty of circular reasoning. Now the data can be partitioned in various ways. Therefore it seems that we can explain *all* the data without being caught into circles. This line of reasoning would certainly be legitimate if the generalization involved would have been inferred from an impressive amount of diverse evidence. For this reason the boundary between epistemic circles and acceptable arguments is not sharp. I would only regard an argument as a plainly unacceptable epistemic circle if, on any reasonable reconstruction, the conclusion represents the only evidence we have in favor of a premise.

The argument in the example could also be criticized on different grounds. The general premise that all green plants do not photosynthesize during the night can be interpreted as a summary of (infinitely many) statements with the form 'species A (a green plant) does not photosynthesize during the night'. Therefore, the premise covertly seems to 'contain' statements about individual species we can infer from it. On this view, we could regard such inferences as logical circles. It is obvious that this implication is unpalatable. If it would be right, all valid deductive arguments would be logical circles! Thus the expression 'logical circle' must be reserved for arguments which contain the conclusion in the premises in a more overt way.

Exercises

3.3.1. A biologist notices that some of the plants he keeps in his room whither away. Inspection shows that there are tiny brown spots with a characteristic shape on the leaves. The spots are familiar to the biologist. They are known to be caused by a virus. Thus, when asked to explain the condition the plants are in, the biologist states that a virus infection explains the phenomenon. Would this amount to an epistemic circle? Notice that it is reasonable to reconstruct the biologist's line of reasoning as a combination of two arguments. First, the spots observed in his specimens (premise) lead him to the conclusion that there is an infection (conclusion). Second, a general thesis concerning the relation between infection and the occurrence of spots (premise) together with the inferred infection in the specimens (premise) explains the spots observed (conclusion).

3.3.2. Evaluate the following argument.

> Two species with similar ecological requirements cannot coexist in the same area. Therefore, two species which coexist in the same area cannot have the same ecological requirements.

3.4. More on Induction

In the preceding sections I have paid more attention to deductive than to inductive arguments since there is no consensus among philosophers on the proper methods of inductive reasoning. Yet we have to deal with induction since it plays a vital role in science. In my opinion the informal methods described by John Stuart Mill in *System of Logic* (1843) still capture much of what happens in science. Three of his methods are outlined below. In the next chapter I consider other methods of induction.

1. Method of agreement. If a factor is always present before a phenomenon of a certain kind occurs, and instances of the phenomenon do not share other antecedents, then this factor is a cause of the phenomenon.

• *Example 3.* Flowering in some plant species that grow in deserts is always preceded by heavy rains. It is reasonable to infer that rain is a cause of flowering. •

• *Example 4.* John drinks lots of lemonade with gin during a party. Afterwards he has a bad headache. Because he does not want this to happen again, he opts for lemonade with scotch at the next party. Again he is plagued by a headache. The result appears to be the same after the consumption of lemonade with rum. John concludes that lemonade is the culprit, and he decides to drink pure liquor in the future.

  This line of reasoning would be acceptable if no further information on drinks would be available. However, the knowledge we do have leads to the rejection of the argument on the ground that it does not comply with the criterion of coherence. •

2. Method of difference. If a factor is always present before a phenomenon of a particular kind occurred, and absent in otherwise similar situations when the phenomenon failed to occur, then this factor is a cause of the phenomenon.

• *Example 5.* In an agricultural area many dead birds are found. It appears that the area has just been sprayed with insecticide. At other times, in the absence of spraying, there was no high mortality among birds. It is reasonable to infer that the mortality was caused by the insecticides. Additional knowledge in this case strengthens the inference. We know that insecticides affect organisms in adverse ways. •

3. Method of concomitant variation. When a factor varies in a positive or a negative way with the degree in which a phenomenon occurs, then there is a causal relationship between the factor and the phenomenon.

• *Example 6.* The observation of a positive relationship between temperature and locomotory activity in a species of beetle supports the assumption that the relationship is causal. •

• *Example 7.* I have been told that there is a positive relationship (a positive correlation) between the salaries of preachers in a particular area in The Netherlands and the export of rum from Jamaica. In this case application of Mill's method would lead one astray. The relationship is an example of a correlation which is spurious for lack of coherence with available knowledge. •

Arguments in the style of Mill can be evaluated with the criteria introduced in section 2. Examples 4 and 7 show that the criterion of coherence deserves special attention here since we are dealing with inductive arguments (see also section 2, example 5).

In the study of spontaneously occurring phenomena, we will seldom be able to implement Mill's methods in a straightforward way. For example, concerning the method of difference we will have to face the problem that the situations compared will normally differ with respect to many factors. How then can we identify a factor which is causally responsible for the phenomenon studied? Mill has two important suggestions which can help us solve this problem. In the first place, we can *restrict the set of candidate-causes by the use of background information.* In the second place, we can *perform experiments in which one factor is manipulated and other factors are kept constant* as far as possible. If Mills methods are put to work in this way they can be quite powerful.

• *Example 8.* In my backyard, indeed throughout the neighborhood where I live, the abundance of birds is limited. In other neighborhoods there are many more birds. The most important difference I can think of concerns cats. Many cats are around where I live; elsewhere there are less of them. It is probable that there will be other differences between neighborhoods which differ in bird abundance. However, in view of background information it is reasonable to infer that cats will be a causal factor. Cats eat birds and birds are afraid of cats. An experiment could provide more confirmation. If I would shoot the cats near my place and bird abundance would subsequently increase, I would feel confident that cats do influence the abundance of birds. However, this is not an experiment I would like to perform.

If an experiment of this kind were indeed performed with positive results (for the birds I mean), the evidence would be telling. However, we should realize that the situations compared—before and after the shooting—may differ in other respects. Thus it is possible that, from a bird's point of view, there happens to be a long-lasting improvement of the weather after the shooting.

In view of this the following experiment would be more decisive. Suppose we identify ten neighborhoods with many cats. We could remove the cats from five randomly chosen neighborhoods, and let the cats be in the remaining ones. If bird abundance would increase in the cat-free areas, not elsewhere, that would be something. It is improbable that the two groups of neighborhoods will systematically differ in another factor that influences birds. •

The identification of causes of a known phenomenon may be regarded as an explanation of the phenomenon. Thus Mill's methods play a role in *inferences to the best explanation* (see Lipton, 1991; he argues that such inferences differ from Mill's methods but come close to them; I will disregard putative differences). In example 8, the aim was to identify a causal factor that would best explain the phenomenon studied.

Lipton argues that *many explanations are contrastive.* Such explanations are a response to a contrastive why-question. The question why some phenomenon *P* occurred thus may be taken to mean, why did *P* occur rather than *Q*? Such a contrastive question is answered by an inductive inference which identifies a difference in causal antecedents between the actual phenomenon *P* and the hypothetical phenomenon *Q*. Example 8 illustrates this. The identification can at once be regarded as an explanation.

Exercises

3.4.1. Penguins mostly occur in areas with low temperatures throughout the year. One could infer from this that the absence of penguins elsewhere is caused by high temperatures. However, penguins appear to thrive in zoos where temperatures are high. Have you any ideas about other causal factors which might be responsible for the distribution of penguins?

3.4.2. At present the level of carbon dioxide in the atmosphere seems to be increasing. Would the following argument be a satisfactory explanation of this phenomenon?

> Devastation of rain forests causes a global increase of the carbon dioxide content of the air; rainforests are now being devastated; therefore, the carbon dioxide content of the air is now increasing.

3.5. Afterthoughts

In chapter *1* I have argued with respect to scientific theories as an example that *the importance of methodological criteria is context-dependent.* Context-dependence likewise surfaced as a theme in chapter *2*. What about the criteria discussed in the present chapter? At first sight validity or inductive support, truth or confirmation, non-circularity, empirical content, and coherence are reasonable criteria which all arguments must satisfy, but appearances are deceptive.

Notice first that these criteria have been tailored to fit the context of empirical science. Thus the criterion of empirical content would be inappropriate in mathematics. Context-dependence also plays a role *within* natural science. I have indeed formulated some criteria in a vague way because the specific form they should take will depend on the context. For example, I would not like to be very specific about the *strength* of inductive support that inductive arguments must exhibit in order to be acceptable. If the truth of an inductively supported conclusion would be a matter of life or death, we should require strong support. In other situations we may be less demanding.

The criterion of coherence is likewise vague. What form should coherence take? Again there is no general answer. Consider an argument in which the occurrence of a storm is inferred from a drop in barometer readings. If the function of such an argument is prediction, it may well be acceptable. It will not do, though, as a causal explanation since barometer readings are not a causal factor in the genesis of storms. We would reject an explanatory argument of this kind since premises concerning relations between barometer behavior and storms would not cohere with scientific knowledge *in an appropriate way.*

I have concentrated on general-purpose criteria which cut across most contexts in natural science. Now one could argue that my list of such criteria is not exhaustive since they do not exclude all unacceptable arguments forms. Two examples come to mind. First, we can infer statements with the form '*p* or *q*' from a premise with the form '*p*', which is weird if the intention is to find support for '*p* or *q*'. Second, we can add whatever premise we want to the premises of a valid argument. This also results in weird arguments, though validity and other criteria are not affected by it.

I would prefer to exclude such arguments by giving the criterion of coherence a broad, intuitive interpretation. The two argument forms would not represent an appropriate kind of coherence. This would obviously leave much to common sense, and it would make coherence an even vaguer notion. Alternatively, you could add additional criteria to my list. Thus a criterion of non-redundancy could serve to exclude arguments in the second category I mentioned (but notice that *3.2*, example 2, suggests that redundant premises may sometimes be acceptable).

All in all I would argue that we should deal with principles of applied logic and methodology in a flexible way. Logic and methodology have a hard core, but its recipes have context-dependent limitations.

CHAPTER 4

Elements of Formal Logic

4.1. Propositional Logic

In the previous chapter I have dealt with logic in an informal way. Thus the evaluation of arguments with respect to logical form was left to intuition. Intuition works well in many cases, but it has its limitations. In the present chapter I will therefore present a brief survey of elementary formal logic, which accounts for logical form in a more rigorous way.

In my experience students of the life sciences react to formal logic in very different ways. Some enjoy it, others detest it. If you belong to the latter category you can skip the appendix to this chapter, which deals with the most technical matters. I do hope that you will come to like elementary formal logic if you don't already. In later chapters I will use the symbolism of formal logic sparingly. Thus they can be understood if the present chapter is skipped or used as a reference guide. However, any formal logic you assimilate will provide you with a deeper understanding.

Deductive logic consists of two parts, *propositional logic* and *predicate logic*. In propositional logic, elementary statements are the basic building blocks. Predicate logic in addition analyses the inner structure of statements. The present section is concerned with propositional logic.

In propositional logic the symbols '*A*', '*B*', '*C*', ... are used as *constants* for particular statements, '*p*', '*q*', '*r*', ... as *variables* for statements. An important role is played by expressions such as 'not' and 'if ... then', *connectives*, which are used to form *compound statements* from *elementary statements*. Thus we get the compound statement 'if *A* then *B*' from the statements '*A*' and '*B*'. Likewise for 'if *p* then *q*', which is a formula rather than a statement. Commonly used connectives, and names for resulting expressions, are presented in the following table.

name	*expression*
negation	not-*p*
disjunction	*p* or *q*
conjunction	*p* and *q*
material implication	if *p* then *q*
material equivalence	*p* if and only if *q* (*p* iff *q*)

The meaning of 'not-*p*' in propositional logic is the same as in ordinary language. It is defined as follows: 'not-*p*' is true if '*p*' is false and *vice versa*.

The expression '*p* and *q*' is likewise unproblematic. It is true if both '*p*' and '*q*' are true, and false in all other cases.

The other connectives correspond less well with ordinary language. In ordinary parlance, '*p* or *q*' is ambiguous. It can mean that *p* is true, or *q* is true, or both are true (inclusive 'or'), or that at most one of them is true (exclusive 'or'). In propositional logic 'or' has the first meaning. So '*p* or *q*' is false only if '*p*' and '*q*' are both false, and true in the remaining cases.

'If *p* then *q*' cannot be rigorously defined in ordinary language. There its truth value (true or false) is undetermined if *p* is false. In propositional logic, the expression is false only if '*p*' is true and '*q*' is false. For the record, the elements '*p*' and '*q*' in the material implication 'if *p* then *q*' have special names; '*p*' is called the *antecedent* and '*q*' the *consequent.*

Lastly, '*p* if and only if *q*' ('*p* iff *q*') is true if '*p*' and '*q*' are both true or both false, and false in the remaining two cases.

Definitions for connectives which do not fit ordinary discourse may strike you as awkward. The material implication is the most telling example. Thus in propositional logic the statement 'If the earth has no moon, then all biologists are stupid' has to be regarded as true because the antecedens (the earth has no moon) is false. In ordinary discourse the statement would be regarded as very odd; some would indeed regard it as false.

In view of such oddities you may wonder why we should use the connectives of formal logic at all. The reason is simply that logic aims to be a rigorous discipline that dispenses with the vagueness and ambiguity that characterize ordinary discourse. For many purposes, not least in science, we need rigor. We have to pay a price for this since the resulting language at times appears to be 'unnatural'. Thus we have to be aware of the meaning we give to connectives in logic. With that proviso the price is well worth paying.

The following example illustrates this with respect to the material implication.

• *Example 1.* The second law of thermodynamics is an important law of physics. In informal terms, it says that the entropy of closed systems never decreases. ('Entropy' means 'amount of disorder', this notion has a precise definition in physics; 'closed system' means 'system which does not exchange energy with the environment'.) In the seventies, there have been heated debates in the scientific literature about the status of this law in biology. Organisms are systems which, in the course of life, show a marked decrease in entropy. Some researchers have argued that they 'contravene' the second law. Others have dismissed this view since organisms are not closed systems so that the law does not apply to them. As the debate continued, a crucial question became whether laws of physics have a restricted validity since they appear to apply only to a restricted set of systems. In my view, the whole discussion was spurious in view of vagueness and ambiguity. The meaning of crucial terms such as 'validity' and 'apply to' in the discussion was anything but clear. The following analysis indicates that an ambiguous use of the expression 'if ... then' may have been an important source of confusion.

The second law can be reformulated as follows: (for all x:) if x is a closed system, then the entropy of x will not decrease. Let's avoid vagueness and ambiguity, and stipulate that the expression 'if ... then' here represents the material implication. The issue now becomes quite clear though it may not fit in with our intuitions; that's a price we have to pay for rigor. In the case of organisms, the antecedent of the law ('x is a closed system') is false, hence the law is 'valid' in the sense of true. (Notice that 'validity' as applied to arguments is a different notion; I am here adopting a terminology that played a role in the discussion.) The law does apply to organisms, organisms do not 'contravene' it. Admittedly, what the law expresses about organisms is a bit trivial. Philosophers would say that it is vacuously or trivially true for organisms. But true it is.

Let me assume that you endorse my search for rigor but that you prefer a different explication of 'if ... then', such that statements containing the expression 'if ... then' are false if their antecedent is false. Thus you can rightly maintain that the second law is *not* 'valid' for organisms. However, this is compatible with *my* thesis that the law is valid for organisms since I use a different connective ('if ... then' with a different meaning) and hence a different concept of validity. Your language is different from mine, so our theses are different even though they are worded in the same way.

A third person could opt for yet another explication and decide that an if-then statement is neither true nor false if its antecedent is false. Thus the law would not 'apply to' organisms since it would say nothing about them. This is compatible with my thesis and with yours. We are merely dealing with linguistic differences.

Throughout the discussion about the law, the facts have not been in dispute. The issue appeared to be a conceptual one. If conceptual matters would have received more attention, we would not have had all this confusion. •

On the basis of the definitions for connectives presented above we can determine conditions for the truth or falsity of compound statements; the procedure for this is explained in the appendix.

Some compound statements will always be true, irrespective of the truth or falsity of their components. Such statements are called *tautologies*. Tautologies are true logical statements (see chapter *1* for the difference between logical and empirical statements, which is important here). 'There will be a thunderstorm tonight or there will not be a thunderstorm tonight' is an example. Notice that you don't need any information on thunderstorms to get at the truth in this case. The form of the statement is '*A* or not-*A*'; *any* statement with this form is true for logical reasons.

Statements which are always false, irrespective of the truth or falsity of their components, are called *contradictions*. Contradictions are false logical statements.

An obvious example is 'There will be a thunderstorm tonight and there will not be a thunderstorm tonight', which has the form '*A* and not-*A*'.

A slightly more difficult example is presented below.

• *Example 2*. A statement with the form 'if (if *A* then *B*) and *A*, then *A*' is a tautology. Let *A* stand for 'John is infected with the HIV-virus' and *B* for 'John will get AIDS'. It is intuitively obvious that the compound statement I formulated is indeed a tautology. It cannot be false because it is logically impossible that the antecedent ('if John is infected ... then he will get AIDS *and* John is infected') is true whereas the consequent ('John will get AIDS') is false. Notice that the negation of the compound statement is a contradiction. •

Two statements are called *inconsistent* if their conjunction generates a contradiction, as in '*A* and not-*A*'. When two statements are not inconsistent they are called *consistent*. Two statements are *equivalent* if connecting them by the iff-connective leads to a tautology. For example, '*A*' is equivalent to 'not-not-*A*', because '*A* iff not-not-*A*' is a tautology.

The validity of argument forms can be assessed according to the following recipe. An argument with premises 'p1, p2, ...' and conclusion 'q' is valid if and only if 'if p1 and p2 and ..., then q' is a tautology. Hence an argument with premises 'p and q' and 'p', and conclusion 'q', is valid.

Notice that this recipe works both ways (cf. 'if and only if'): we can move from a tautology to a valid argument and *vice versa*. You should realize also that the recipe applies to cases with a single premise, and that it does not preclude arguments in which a premise and the conclusion are identical. The recipe does apply quite generally.

Let's first consider the simplest case conceivable. 'If there will be a thunderstorm tonight, then there will be a thunderstorm tonight' is a tautology with the 'if ... then' connective. Therefore an argument with the single premise 'There will be a thunderstorm tonight' and the conclusion 'There will be a thunderstorm tonight' is valid. That's trivial of course.

If we apply the recipe to example 2 we get the following result.

• *Example 3.* We can show that the following argument is valid. Premise 1. If John is infected with the HIV-virus then he will get AIDS. Premise 2. John is infected with the HIV-virus. Conclusion. John will get AIDS. •

Some valid and fallacious argument forms are presented below in accordance with a common notational scheme with premises and conclusion separated by a horizontal bar.

valid argument forms

$$\frac{\text{if } p \text{ then } q}{q}$$ modus ponens

$$\frac{\begin{array}{c}\text{if } p \text{ then } q\\ \text{not-}q\end{array}}{\text{not-}p}$$ modus tollens

$$\frac{\begin{array}{c}p \text{ or } q\\ \text{not-}p\end{array}}{q}$$ disjunctive syllogism

$$\frac{\begin{array}{c}\text{if } p \text{ then } q\\ \text{if } q \text{ then } r\end{array}}{\text{if } p \text{ then } r}$$ hypothetical syllogism

fallacious argument forms

$$\frac{\begin{array}{c}\text{if } p \text{ then } q\\ q\end{array}}{p}$$ fallacy of affirming the consequent

$$\frac{\begin{array}{c}\text{if } p \text{ then } q\\ \text{not-}p\end{array}}{\text{not-}q}$$ fallacy of denying the antecedent

$$\frac{\begin{array}{c}p \text{ or } q\\ p\end{array}}{\text{not-}q}$$ fallacy of asserting an alternative

• *Example 4.* 'If evolutionary theory is true, then the fossil record will always show gradual transitions. The fossil record does not always show gradual transitions. Therefore evolutionary theory is not true.' This is a valid deductive argument with the form of the modus tollens. If the argument is not accepted by biologists, the reason will not be that it is invalid, but rather that the first premise is regarded as false. •

Examples 3 and 4 are rather straightforward in that the analysis of validity they provide accords well with intuition. Now the notion of validity is linked up with the material implication ('if ... then', see above), which has some counterintuitive feautures (see example 1). Hence we may also expect some counterintuitive results with validity. Examples 5 and 6 confirm this. That is a price we must pay for a rigorous concept of validity. I will indicate in the examples how we can accommodate counterintuitive results.

• *Example 5.* You might not guess that the following argument is valid, but it is. 'It is raining today and it is not raining today. Therefore, all biological theories are false.' The argument is valid because the sentence form 'if p and not-p, then q' represents a tautology according to logical convention. Thus we can derive any statement whatsoever from a contradiction.

This shows that the relationship 'follows from' in logic does not cover all the shades of meaning the expression has in ordinary discourse. In this respect 'follows from' in logic takes over the behavior of 'if ... then' (see above). In ordinary discourse we would call the argument invalid because it is absurd. It is impossible to capture ordinary meanings by any formal system in an exhaustive way. This is not a cause for despair. Logic (or, more broadly, methodology) does take care of the absurdity, but it gives it a different locus: the argument is not acceptable since it is impossible that the premise is true. •

• *Example 6.* The following argument is as strange as the one in example 5. 'If the earth has a moon, there are elephants in Africa. The earth has a moon. Therefore, there are elephants in Africa.' The argument is valid, and its premises are true. Yet we would not regard it as acceptable, because the first premise does not satisfy the criterion of coherence discussed in 3.2. This is again an example which shows that the language of logic deviates from ordinary discourse. We would normally regard the first premise as odd or even false. The 'if ... then' of the material implication in logic obviously does not cover all the connotations the expression has in ordinary discourse. We should not conclude from this that the tools of logic are inappropriate. Deviations from ordinary language are unavoidable if we aim at rigor. In the present example some aspects of meaning which are lost in the process of translation can be taken into account by the criterion of coherence. According to this criterion the argument is unacceptable. •

Exercises

4.1.1. In order to be acceptable, an argument must have consistent premises. Show that this criterion follows from criteria introduced in chapter *3*.

4.1.2. Consider the following argument.

> If the population density of a species is high in some area, the species will not reproduce in that area. If a species does not reproduce in some area, it will go extinct in that area. Therefore, if the population density of a species is high in some area, it will go extinct in that area.

What is the logical form of this argument? Is the argument valid? Would a biologist accept the argument?

4.1.3. 'Worldwide there cannot be much air pollution, because lichens are still common.' Someone who utters this statement presumably means it as an argument. Reconstruct the statement in such a way that the form of the argument becomes clear.

4.2. Predicate Logic

In propositional logic elementary statements are the building blocks. Predicate logic in addition considers the internal structure of such statements. In chapter *3* predicate logic was already used in an informal way; the example in *3.1* covers the spirit in which it deals with logical structure in a preliminary way.

If you want to assess arguments in science, a keen awareness of the way statements are structured is often indispensable. We must realize that there is no such thing as the structure of a statement. Any statement can be reconstructed in different ways. The purposes you have with an analysis will determine which reconstruction is adequate. Note that all this confronts us again with the theme of context-dependence, which plays a role throughout this book.

The issue of reconstruction applies also to the global structure of arguments as studied by propositional logic; for ease of exposition I have not made it a theme in the previous section. You should also realize that propositional logic and predicate logic may provide alternative reconstructions of an argument which suit different purposes. One of *my* purposes in the previous section was to acquaint you with the notion of validity. For that purpose, I reconstructed some arguments in a minimal way. I stayed with propositional logic and did not comment on the structure of statements. If my purpose would have been to present a full-fledged evaluation of the arguments the reconstruction would have been totally inadequate.

Predicate logic is a means, an important one, to come to grips with the structure of statements and arguments. Even within predicate logic, many alternative reconstructions are always possible. In view of this it is advisable to stick to the following rule: *do never use a complex reconstruction if a simple one suits your purposes as well.*

Complicated expressions can often be taken to stand for a single property. Consider the statement that John has a good salary which allows him to buy cars all the time. The simplest way to reconstruct this is as follows. We can stipulate that 'has a good salary which allows one to buy cars all the time' expresses a single property. Let us denote it by the symbol 'P'. Thus we get 'John has property P'. Further, we can introduce a symbol, say 'a', for the individual John. In predicate logic the result of all this ('a has property P') is conveniently expressed as 'Pa'.

The following symbols are used in predicate logic. 'P', 'Q', 'R', ... are *constants* standing for properties of things; 'a', 'b', 'c', ... denote individual things; 'x', 'y', 'z', ... are *variables* for things. The statement that thing a has property P is expressed by 'Pa'. This statement *instantiates* the expression 'Px', which represents the category of such statements. Relations are covered as follows. 'Rab' expresses the statement that a relationship R exists between a and b. 'P' in the expression 'Pa' is a *one-place predicate*, 'R' in 'Rab' is a *two-place predicate* (see 2.5).

In addition to this the following expressions are used in predicate logic:
The formula '$(x)(... x ...)$' means 'for all x, ... x ...'. The symbol '(x)' represents the *universal quantifier*; a statement with the form of the formula is a *universal statement.*
The formula '$(\exists x)(... x ...)$' means 'there is at least one x, such that ... x ...'. The symbol '$(\exists x)$' represents the *existential quantifier*; a statement with the form of the formula is an *existential statement.*
Notice that 'Px' is a statement *form*, not a statement, since it does not ascribe property P to any particular thing. In contrast to this, the expressions '$(x)(Px)$' and '$(\exists x)(Px)$' are full-fledged statements. The quantifiers in such statements are said to *bind* the variable in the expression 'Px'.
A statement which does not contain a quantifier is called a *singular statement.*

The example below indicates how *statements* are *translated from ordinary discourse into* the idiom of *predicate logic.*

• *Example 1.* The statement 'John is infected with the HIV-virus' ascribes a particular property, 'is infected with the HIV-virus' to an individual. If we use the symbol '*P*' for this property, and '*a*' for the individual, John, we get '*Pa*'. Notice that I am using the term 'property' in a wide sense. In predicate logic, properties need by no means be simple features. The statement that John will get AIDS, analogously, can be represented as '*Qa*', in which '*Q*' stands for 'will get AIDS'. Hence we get 'if *Pa* then *Qa*' for the compound statement that John will get AIDS if he is infected with the HIV-virus. The statement that all persons who are infected with the AIDS-virus will get AIDS thus becomes '(*x*)(if *Px* then *Qx*)'. If it is phrased in this way we should add that '*x*' is a variable which only ranges over persons. Alternatively, and more neatly, we can decide not to restrict the variable in this way, and let '*P*' stand for 'is *a person* infected ...' and '*Q*' for '*is a person who* will get AIDS'. •

Intuition will often suffice to evaluate arguments in predicate logic, particularly if parallels with propositional logic are considered (possibilities for a more rigorous evaluation are outlined in the appendix). The following argument is obviously valid since it represents the modus ponens.

$$\frac{\text{if } Pa \text{ then } Qa}{Qa}$$

The argument below, which is also valid, is related to it.

$$\frac{(x)(\text{if } Px \text{ then } Qx)}{Qa}$$

The next example makes this more concrete.

• *Example 2.* Suppose John has indeed developed AIDS. Why did he get AIDS? Let's regards this as a question that calls for a scientific explanation. An answer we can give is: John developed AIDS because he was infected with the HIV-virus. Thus the explanation takes the form of a statement. It is presumably meant, though, as an argument. If we use the same symbols as in example 1, we can give the explanation the following form.

$$\frac{\text{if } Pa \text{ then } Qa}{Qa}$$

However, this is not quite satisfactory. Anyone who explains John's getting AIDS in terms of HIV-infection, will doubtless be thinking that he got AIDS because HIV-infection leads to AIDS *in general*. Hence the explanation is more appropriately construed as follows.

$$(x)(\text{if } Px \text{ then } Qx)$$
$$\frac{Pa}{Qa}$$

This reconstruction shows that the logic of arguments may be much more complex than formulations in ordinary discourse suggest. •

The following example is more complicated because it involves two-place predicates.

• *Example 3.* The argument in *3.2*, example 1, about the relation between fitness and survival, has the following form.

$$(x)(y)(\text{if } Sxy \text{ and } Fxy, \text{ then } Cxy)$$
$$\frac{Sab \text{ and } Fab}{Cab}$$

The expressions '*Sxy*', '*Fxy*', and '*Cxy*' stand for 'individual x and individual y belong to the same species', 'individual x has a higher fitness than individual y', 'individual x has better chances of survival than individual y', respectively. The argument is valid. •

In ordinary discourse, quantifiers often remain implicit. It is important that we be aware of them. Suppose you would hear somebody saying that politicians are liars. Would you agree? I don't mean what opinion you would voice by way of a response in small talk. Let's be serious about this. Before we answer such a question we should ask how it is intended. One interpretation is that all politicians are liars all the time; thus we get a statement with two universal quantifiers. On most occasions this will not be a plausible interpretation of the statement. On another interpretation we get two existential quantifiers. This yields a plausible version ('there is at least *one* ...') which is easily tested, but it is not very informative (how many?).

Quantifiers are connected with testability in important ways. The connections are worked out more fully in chapter 6. As a prelude to this theme let me briefly introduce two species of testability, verifiability and falsifiability. A statement is verifiable if it is possible (logically so) to show that it is true *if* it is true. Likewise for falsifiability (replace 'true' by 'false'). Back to the politicians. Our two interpretations clearly differ in testability because they have different quantifiers. It would be difficult to verify 'All politicians ...', but one could easily falsify it. The converse holds for 'There is at least one politician ...'. In brief, universal statement are falsifiable, not verifiable; existential ones are verifiable, not falsifiable (complications and exceptions will be discussed in chapter *6*).

The next example is an additional illustration of the connection between quantifiers and testability.

• *Example 4*. In ordinary discourse we will say that persons who are infected with the HIV-virus get AIDS, if what we really mean is that *all* rather than *some* infected persons will get the disease. Needless to say, it will be less easy to verify a universal statement than to verify an existential one concerning the relationship between the virus infection and the disease. Even if all the evidence we have confirms a universal statement, it remains possible that a counter-example will turn up. Hence we can never be certain that such a statement is true. With respect to existential statements the situation is different. Evidence in the form of one positive instance will suffice to verify such a statement, provided that the evidence does not depend on assumptions which also need to be tested. For that matter, such assumptions almost always play a role in scientific work. •

A positive universal statement with the form '$(x)(Px)$' is equivalent to a negative existential statement with the form 'not-$(\exists x)$(not-Px)'. In words: 'for all x, x has property P' has the same meaning as 'there is no x which does not have property P'. Hence the remarks I made on testability must be qualified. They apply only to positive statements. The table below indicates what equivalence relations there are between statements containing a universal or an existential quantifier. In all the four cases represented, the expression at the left-hand side is equivalent to that at the right-hand side.

$$(x)(Px) \qquad \text{not-}(\exists x)(\text{not-}Px)$$
$$(\exists x)(Px) \qquad \text{not-}(x)(\text{not-}Px)$$
$$(x)(\text{not-}Px) \qquad \text{not-}(\exists x)(Px)$$
$$(\exists x)(\text{not-}Px) \qquad \text{not-}(x)(Px)$$

In addition to universal statements and existential ones there are those which contain both a universal and an existential quantifier. 'All people have friends' is an example. We could reconstruct it as a purely universal statement ('All people have the property of having friends') but for most purposes that would not be very adequate. The logical structure of the statement is better revealed by the following reconstruction: 'For all people, *there is* at least one person who is a friend', which contains an existential quantifier in addition to the universal one. Such a 'mixed' statement, according to the arguments set out above, is neither verifiable nor falsifiable. This is indeed so in this case if we use the notions of verifiability and falsifiability in the strict sense they are intended to have in philosophy. (Think about how you should proceed to find out with absolute certainty if the statement is true or false!) We all think, though, that the statement is false.

Rightly so, since *it is legitimate* in this case *to take a great variety of background information for granted.*

In science, important theses may covertly contain both a universal and an existential quantifier. We should be aware of this in view of consequences for testability. As the next example shows, uncovering quantifiers may have unexpected results that affect research.

• *Example 5.* Consider the thesis that all features of organisms have a function. One way to test this thesis is by searching for a counter-example. At first sight the thesis is falsifiable because the finding of a counter-example will prove the thesis false. But could we prove in any particular case that we have uncovered a counter-example? Here we are in for a surprise. Logical reconstruction shows that it is impossible to find evidence that is absolutely convincing.

The logical form of the thesis is rather complex. One way to express it is as follows:

$$(x)(\text{if } Px, \text{ then } (\exists y)(Qy \text{ and } Rxy))$$

Here 'P' stands for 'is a feature of an organism', 'Q' for 'is a function', and 'Rxy' for 'x has function y'. In words the expression means: 'for all x, if x is a feature of an organism, than there is a function which x has'. If we decide to use 'x' as a variable for features of organisms and 'y' as a variable for functions, we can also use a simpler expression:

$$(x)(\exists y)(Rxy)$$

The important point of this reconstruction is that it uncovers a universal *and* an existential quantifier. This implies that it would not be easy to demonstrate that there is a counter-example. In the face of a putative counter-example someone can always argue that *there is* a function of the feature studied which remains to be discovered. Those who are unaware of the presence of an existential quantifier might uncritically accept inconclusive data as evidence against the thesis. •

Exercises

4.2.1. Suppose that no stinging-nettles (a species of plant) are present in a particular area. A biologist explains this by noting that the soil in the area has a low nitrogen content. Reconstruct the explanation with the help of predicate logic.

4.2.2. In *3.3*, example 1, the following argument was considered.

The theory of evolution does not allow the derivation of predictions. Theories are adequate only if they allow the derivation of predictions. Therefore, the theory of evolution is not an adequate theory.

Read the example again before answering the following question.

Do you agree that the following reconstruction captures the form of the argument?

$$(x)(\text{if } Tx \text{ then } Px)$$
$$\frac{\text{not-}Pa}{\text{not-}Ta}$$

Here '*T*' means 'is an adequate theory', '*P*' means 'allows predictions', '*a*' denotes the theory of evolution.

4.3. Inductive Logic

In *3.4* I introduced Mill's methods of inductive reasoning, which play an important role in science. In line with common practice, I have presented his methods in an informal way. It would be difficult to formalize them. His methods by no means exhaust possibilities of inductive reasoning. In this section I discuss another important method, which is more easily formalized. Additional comments on inductive arguments are presented in the appendix.

People who are heavy smokers have an increased risk of contracting cancer of the lung. We could specify the risk in the premise of an argument designed to show that some heavy smoker has a particular probability of getting lung cancer. Such an argument would be a *probabilistic inductive argument* with the following form.

$$p(L|S) = q$$
$$\frac{Sa}{La} \quad [q]$$

The first premise is a probabilistic statement expressing the *conditional probability* that feature L will be present given that S is present. It says that the probability of L, given S, is q. '*L*' stands for 'getting lung cancer', '*S*' for 'being a heavy smoker'. The value taken by '*q*' is a statistical probability. So q takes values between 0 and 1 only. The expression '$[q]$' stands for a so-called inductive probability. It covers a probability relation between premises and conclusion.

Notice that it would be unreasonable to demand that the premises of probabilistic arguments be exactly true. Thus the first premise of the probabilistic argument presented above will be acceptable if there are good grounds for assuming that the value of q is close to the actual value.

Many researchers would regard the argument as acceptable only if the value of q is rather high, perhaps close to one. This view is right if one means by it that the premises do not warrant the acceptance of the conclusion unless the value of q is sufficiently high. In the smoking example the value of q will not be so high that the premises warrant the acceptance of the conclusion. The argument as it stands, therefore, is not very satisfactory.

I want to digress a bit about the smoking example because it links up with the theme of *contrastive explanation* which was briefly discussed in *3.4* (I will return to it in chapter *8*). You may feel that smoking does explain the occurrence of lung cancer in some cases since it is reasonable to identify smoking as one of the causes. (If you argue in this way, your line of reasoning will not be fully covered by the argument set out above. I will not elaborate this.)

Should the identification of a cause count as an explanation? I will argue that this depends on the question we are asking. Apart from that the issue is to some extent a terminological one since *the concept of explanation can be used in broad or in narrow senses.*

Suppose we ask why John, a heavy smoker, contracted lung cancer. This question can be interpreted in various ways. Maybe we want to know why John got the disease whereas Peter stayed healthy. If Peter was also a heavy smoker, reference to smoking as a causal factor will not be explanatory. If Peter did never smoke, smoking will be relevant because it is a causal factor which 'makes a difference'. Likewise for the question why John contracted the disease instead of staying healthy.

On all the interpretations I mentioned, the question asked is contrastive. The explanation given in response to such a question may be called a contrastive explanation.

Probabilistic arguments, like other inductive ones, can be overturned by relevant information not covered by the premises. This is illustrated by the following example.

• *Example 1.* Suppose Mary gets Pfeiffer's disease. It appears that she has been in close contact with a friend who suffered from the disease. From this one may infer that the friend has infected her. Transmission of the virus responsible for the disease is known to occur, though it does not happen easily. It is also known that someone will get Pfeiffer's disease only if they are in a poor condition. Indeed Mary had not been feeling well for a long time.

Mary's contracting the disease could be covered by the following argument, in which 'P' stands for 'has Pfeiffer's disease', 'D' means 'has been in close contact with a person with the disease', and 'a' denotes Mary.

$$p(P|D) = q$$
$$\frac{Da}{Pa} \quad [q]$$

This argument will not have much force because the value of q will be low. Also, the information that Mary's condition was poor is left out. According to the criterion of coherence (see *3.2*) the following argument, in which 'Q' stands for the poor condition, is more adequate.

$$p(P|D \text{ and } Q) = r$$
$$\frac{Da \text{ and } Qa}{Pa} \quad [r]$$

The force of this argument is still limited since the value of r will still be low, though it will be higher than that of q. •

In the probabilistic arguments discussed in the last example the inference goes from cause to effect. We could also reason from effects to causes. Concerning the example we can reasonably argue that two causes of a phenomenon (Mary's disease) were identified. This may count as an explanation in a broad sense of the term.

We could obviously continue the search for causes along the lines set out in the example. This would indeed lead to an unlimited quest. *Never would we arrive at an explanation which is complete in the sense that all possible causes are covered. In practice that is certainly impossible; I would also regard it as a matter of principle. Hence explanations have limitations which need not be shortcomings.*

This incompleteness problem implies that reference to causal conditions which are not specified is often appropriate in science. We should be aware of this, especially so because the role of such conditions often remains implicit in scientific writings.

To illustrate the role of conditions which are not specified I will reconstruct the information in the foregoing example in a different way. We could argue that, in addition to the causal factors identified, other causal conditions, not known to us, must have played a role simply because the phenomenon we are aiming to explain did occur. As the following example shows, *a probabilistic argument can be transformed into a 'sketch' of a deductive argument* by reference to such conditions.

• *Example 2.* Let C_i stand for the remaining, unknown causal conditions that played a role in Mary's getting Pfeiffer's disease (see example 1). The assumption that there are such conditions leads to the following valid deductive argument.

$$(x)(\text{if } Dx \text{ and } Qx \text{ and } C_i, \text{ then } Px)$$
$$Da \text{ and } Qa$$
$$\underline{C_i}$$
$$Pa$$

Some would regard this argument as a fallacy, viz. an epistemic circle. As argued in *3.3* we must be cautious with this label. The conclusion *Pa* was indeed used to infer the premise C_i. If Pa would be the sole evidence for C_i the argument would be problematic. However, the inference may be reasonable if there is additional evidence in the form of data or plausible assumptions. In the Pfeiffer case it is doubtful whether such evidence exists. •

Exercise

4.3. A certain species of frog is common in ponds in a particular area. On the basis of an inventory it is estimated that the frog is present in 95% of the ponds which contain *Hydrocharis morsus-ranae*, a species of water plant. The authorities in the area have been somewhat careless with waste disposal. As a result the water in some ponds is polluted with detergents. The frog is present in 10% of the contaminated pools which have been investigated. Suppose that a pond which has not been investigated so far contains both *Hydrocharis* and the detergent.

If no more information is at your disposal, would you infer that the frog occurs in this pond? Would your conclusion be different if you are allowed to take biological theories into account?

4.4. Asides on Causation

In the previous section I have freely used the concept of causation without giving an explication. By way of an apology, let me note that the concept is tricky. There is no consensus in philosophy about any particular explication. As explained below, this is due in part to the fact that the category of causes is heterogeneous. I will give a partial explication of various meanings of 'cause'.

The notions of *necessary condition*, *sufficient condition* and *necessary-and-sufficient condition* are important for understanding causation. I will explain them with the help of the formalism of propositional logic and predicate logic.

If some statement 'if *p* then *q*' is true, the state of affairs described by '*p*' is sufficient condition for the state of affairs described by '*q*'. That is, if *p* is realized, *q* cannot fail to occur. The converse is not true, but *q* is a necessary condition for *p*. If *q* does not occur, then *p* does not occur either. The statement '*p* iff *q*' represents a necessary-and-sufficient condition which goes both ways, because it implies both 'if *p* then *q*' and 'if *q* then *p*'.

Analogously, one can say that feature *P* is a sufficient condition for feature *Q* if the statement '(*x*)(if *Px* then *Qx*)' is true, and so forth.

For many practical purposes, elementary logic can be used to express conditions which are *causally* necessary, or sufficient, or necessary-and-sufficient, although the full strength of these notions is not captured by it. *We should remain aware of connotations which get lost when we express causal idiom in this way.* In ordinary language, we will say that a condition causes an effect, if the effect can be changed by manipulating the condition. Also, effects are mostly assumed not to precede causes. These aspects of causal relationships are not covered by the expressions I used.

The term 'cause' is sometimes used for necessary conditions, but it may also stand for sufficient or necessary-and-sufficient conditions. In addition to this there are causes in a weaker sense. Smoking is a cause of lung cancer, but it is neither necessary nor sufficient for it.

Mill's methods aim at the detection of causes in particular senses of the term. At first sight, the method of agreement, as formulated above, simply envisages causes as necessary conditions, whereas the method of difference aims to identify factors which are causally necessary-and-sufficient. However, this must be qualified.

Let's reconsider section *3.4*, example *8*. Here the method of difference was applied to data concerning bird abundance in particular neighborhoods. For this purpose neighborhoods with cats were compared with those without cats. Application of the method led to the conclusion that the presence of many cats is a cause of low bird abundance *in the relevant neighborhoods*. Now cats may be regarded as a necessary-and-sufficient cause if the effect is formulated in this specific way. Needless to say, cats are by no means a necessary-and-sufficient cause of low bird abundance *in general*.

All this shows that we must carefully specify the things we are comparing. I will illustrate the importance of this by reconsidering example 2 of *2.5*. As argued in that section, the concept of genetic determination must not be applied to features of organisms *simpliciter*. All features of organisms are determined by genetic *and* environmental factors. So the distinction of genetically determined and environmentally determined features is problematic. The concept should rather be applied to differences between organisms in a feature. Differences in skin color between people belonging to different races may be genetically determined (be due to a genetic difference, not an environmental difference). Other differences in skin color may have an environmental cause. Many, perhaps most, differences between organisms are due to genetic *and* environmental differences.

The causal factors we identify can obviously be different under different comparisons. Research in biology and medicine does not always take this into account. This leads to problematic generalizations.

For example, it is often claimed that severe psychiatric disorders are genetically determined. Such claims are based on particular comparisons of ill persons with healthy ones. If a particular genetic difference is found, the disorder is said to be genetic, even if it is known that there are environmental differences as well. We should realize that the genetic difference may well be absent in other comparisons, say, in other cultures.

4.5. Afterthoughts

The principles of elementary logic do not always accord with the way we reason in daily life and in science. Indeed no formal system of logic is able to capture the precise meaning of vital notions such as 'if ... then ...' and 'follows from' in ordinary discourse.

Specifically, the use of the material implication in ordinary logic is a source of trouble. It leads to the attribution of truth to statements we would ordinarily regard as odd or even false. The concept of validity of ordinary logic, analogously, applies to arguments we would normally regard as incoherent or invalid.

We could try to develop more sophisticated systems of logic which come closer to common modes of reasoning. Various systems aiming at this do exist. However, *total harmony with common discourse in science has not been achieved.*

I have opted for the most elementary system of logic because this book is meant as a textbook. In my view the difficulties I mentioned need not really be troublesome. I have taken them into account by introducing *a broad methodological criterion of coherence* (see *3.2*), which I have not spelt out in formal detail, deliberately so. This criterion should lead to the rejection of arguments that we regard as inadequate despite their passing the net of standards formulated in logic.

Consider the following example. From the premise that AIDS is caused by a virus if the atmosphere contains oxygen, and the premise that the atmosphere does contain oxygen, we can infer that AIDS is caused by a virus. If this example is made to fit the formalism of propositional logic, we will get a valid argument with true premises. In the process of translation, something gets lost. In ordinary discourse, we would not simply say that the first premise is true. A natural reaction would be to regard it as so absurd that the concepts of truth and falsity do not apply. The very rigor of propositional logic (a virtue needed on many occasions) precludes the assignment of absurdity instead of truth or falsity to the first premise. Hence we must part company with ordinary language to some extent. It remains possible, though, to take account of the absurdity. According to the criterion of coherence the first premise is unacceptable since it fails to cohere with available knowledge.

In science we need the formal rigor of logic and mathematics. We also need ordinary discourse. There is a tension between the two which we cannot totally eliminate. We often need to translate ordinary language of science into a formal idiom which permits a rigorous manipulation of data and ideas. After the manipulation we will translate results back to ordinary language. In the translation processes we will always lose and distort things to some extent. That is something we can only account for in informal ways. At the very least all this will help us clarify ideas that have not been articulated yet in satisfactory ways.

The concept of *causation* is an appropriate example. Philosophers have not quite managed to translate ordinary causal language into a formal idiom which is totally appropriate. I have rested content with a relatively primitive translation which relies on propositional logic and predicate logic. Under this translation, various shades of meaning are lost, so we must be aware of them in the evaluation of translated causal arguments. Not everything is lost, though, and we can surely use elementary logic to evaluate causal arguments.

Concerning the concept of causation an important point is that it represents a heterogeneous category of phenomena. As I will argue in later chapters the same is true of other crucial notions in the philosophy of science, for example 'scientific theory' and 'scientific explanation'. Because of this I would argue that it is futile to search for the philosophical model of causation, scientific theories, scientific explanation, whatever. Philosophical models will always have limitations. This need not stand in the way of applications. We should take the limitations into account by applying models in contexts where limitations are relatively harmless, or by supplementing philosophical analyses based on models with a proper dosage of common sense (Sloep and Van der Steen, 1988).

One could object that common sense is a poor guide which needs to be replaced by more rigorous standards as developed in logic and philosophy. I would not agree. Those who try to improve on extant models of logic and philosophy as applied to science, do so in view of counter-examples against these models. Now a counter-example can be declared a counter-example only on the assumption that it represents good science not covered by the model criticized. This assumption is obviously a matter of common sense. It implies that one knows, at the very least in a tacit way, under what conditions extant models are applicable.

APPENDIX TO CHAPTER 4

4.A. Propositional logic

In *4.1* I have kept the use of symbols at a minimum to make formal logic as accessible as possible. A disadvantage of this is that the use of words for connectives may suggest that they are ambiguous, as in ordinary discourse. To prevent confusion, the connectives are represented by symbols in logic, as indicated in the following table.

name	expression	symbol
negation	not-p	$\neg p$
disjunction	p or q	$p \vee q$
conjunction	p and q	$p \wedge q$
material implication	if p then q	$p \rightarrow q$
material equivalence	p if and only if q	$p \leftrightarrow q$

Definitions for connectives are summarized in the following tables, in which 'T' stands for 'true' and 'F' for 'false'. The tables are known as the *fundamental truth tables*.

p	$\neg p$
T	F
F	T

p	q	$p \vee q$	$p \wedge q$	$p \rightarrow q$	$p \leftrightarrow q$
T	T	T	T	T	T
T	F	T	F	F	F
F	T	T	F	T	F
F	F	F	F	T	T

The truth conditions for more complex formulas can be determined by successive application of the fundamental truth tables.

The following table, which is a *derived truth table*, illustrates this.

p	q	$p{\to}q$	$(p{\to}q){\wedge}p$	$((p{\to}q){\wedge}p){\to}q$	$\neg((p{\to}q){\wedge}p){\to}q)$
T	T	T	T	T	F
T	F	F	F	T	F
F	T	T	F	T	F
F	F	T	F	T	F

The third column in this table is simply the definition of '$p{\to}q$'. We can proceed with the next column by applying the definition of '$p{\wedge}q$' to '$p{\to}q$' and 'p' (instead of 'p' and 'q') as units. And so forth.

Notice that we need to use brackets in complex compound formulas or statements to avoid ambiguities. The number of brackets can be reduced by observing the following rule for the binding force of connectives. The binding force decreases in strength in the series $\neg \wedge \vee \to \leftrightarrow$. Thus the formula '$((p{\to}q){\wedge}p){\to}q$' reduces to '$(p{\to}q){\wedge}p{\to}q$'.

The last two formulas in the derived truth table will always generate true and false statements, respectively. The truth or falsity of their components does not make a difference. These formula therefore stand for logical statements, *tautologies* and *contradictions*, respectively.

An argument with premises '$p1, p2, ...$' and conclusion 'q' is valid if and only if '$p1{\wedge}p2{\wedge}...{\to}q$' is a tautology. Hence an argument with premises '$p{\to}q$' and 'p', and conclusion 'q', is valid. I will not reproduce the list of valid and fallacious argument forms. It is easy to cast the original list in a more appropriate form; one only needs to replace words representing connectives by appropriate symbols.

4.B. Predicate logic

Predicate logic does not have an easy procedure like the truth table method to evaluate statements and arguments associated with them, but there are more indirect ways to do this. We can obtain tautologies in predicate logic by a procedure illustrated in the following example. The formula '$(p{\to}q){\wedge}p{\to}q$' stands for a tautology in propositional logic. If we substitute 'Px' for 'p' and 'Qy' for 'q' we get a tautological formula in predicate logic. If we bind the variables with universal quantifiers, we get the following tautological statement: '$(x)(y)((Px{\to}Qy){\wedge}Px{\to}Qy)$'. The statement '$(x)((Px{\to}Qx){\wedge}Px{\to}Qx)$' is likewise a tautology.

Predicate logic presupposes the validity of the following argument.

$$\frac{(x)(Px)}{Pa}$$

Since we can substitute more complex expressions for '*Px*', the following argument is also valid:

$$\frac{(x)(Px{\rightarrow}Qx)}{Pa{\rightarrow}Qa}$$

Now the following argument is valid since it represents the modus ponens in propositional logic.

$$\frac{\begin{array}{c}Pa{\rightarrow}Qa\\Pa\end{array}}{Q}$$

So the following argument is also valid:

$$\frac{\begin{array}{c}(x)(Px{\rightarrow}Qx)\\Pa\end{array}}{Qa}$$

Some important tautologies which we cannot derive from propositional logic are presented below (cf. the table presented in *4.2*).

$$(x)(Px) \;\leftrightarrow\; \neg(\exists x)(\neg Px)$$
$$(\exists x)(Px) \;\leftrightarrow\; \neg(x)(\neg Px)$$
$$(x)(\neg Px) \;\leftrightarrow\; \neg(\exists x)(Px)$$
$$(\exists x)(\neg Px) \;\leftrightarrow\; \neg(x)(Px)$$

Common sense confirms the tautological nature of these expressions. Thus the thesis that all things have a particular property (left hand side of the first expression) is equivalent to the thesis that there is no thing which does not have this property (right hand side of the first expression).

4.C. Inductive logic

In *3.4* and in *4.3* I have presented some important inductive argument forms. For the record, I will discuss one additional, elementary inductive argument form in this section. Its usefulness is limited. That's why I did not discuss it in the main text.

Inductive enumeration represents the simplest kind of inductive reasoning. Here is an example.

$$Sa \land Wa$$
$$Sb \land Wb$$
$$Sc \land Wc$$
$$\dots\dots$$
$$\overline{\overline{\dots\dots}}$$
$$(x)(Sx \rightarrow Wx)$$

'Sa' means 'a is a swan', 'Wa' means 'a is white'. Thus I have moved from statements expressing the whiteness of individual swans to the statement that all swans are white. The double line indicates an inductive relationship. My presentation seems a bit inappropriate because the conclusion has a different connective than the premises. From a logical point of view, one would expect premises with the form '$Sa \rightarrow Wa$' ... (which is implied by '$Sa \land Wa$' ...) or a conclusion with the form '$(x)(Sx \land Wx)$'.

The latter reconstruction will not do because its (plainly false) conclusion is obviously not the one we would like to derive. The former one is unsatisfactory for a different reason. Expressions with the form '$Sx \rightarrow Wx$' would be true for white swans, but also for all things which are not swans, for example black shoes. Yet the observation of such things would hardly support the intended meaning of the conclusion (in the formal presentation the conclusion does cover black shoes). The use of the material implication apparently fails to capture the fact that the argument is supposed to be only about swans. Thus we need the criterion of coherence (see 3.2) to exclude statements with the form '$Sx \rightarrow Wx$' from the premises.

The criterion of coherence for inductive arguments also demands that all relevant information is included in the premises. Therefore the above argument is overruled by the observation of a black swan. Notice that the premises would also support the conclusion '$(x)(Wx \rightarrow Sx)$' on the assumption that all relevant information is included. Needless to say, the assumption is manifestly false in this case.

CHAPTER 5

Scientific Research: An Overview

Science is often characterized as the hallmark of reliable empirical knowledge. To the extent that this title is deserved it is due to the methods by which knowledge is produced. Many philosophers and scientists hold that the strength of scientific method does not reside in the way new *ideas* about reality are generated. The point is rather that such ideas, once we have them, are subjected to severe *tests*.

Generality is an additional feature which supposedly characterizes science. A compilation of sundry facts about a forest, a town, or American presidents, however reliable, is not science. Scientific knowledge typically takes the form of general theories.

Accordingly we get the following picture of scientific research. Tentative ideas about phenomena scientists are interested in are called *hypotheses*. General empirical statements, however outrageous, may be acceptable as hypotheses provided we are willing to test them. *We can arrive at hypotheses in many different ways.* Arduous fact gathering, the study of scientific literature, flashes of insight we get in dreams, any activity whatsoever may lead to valuable ideas.

Because hypotheses are general we cannot simply infer them from statements about facts by a process of deduction. Facts at best provide inductive support. However, there is a powerful role for deduction in critical *tests* of hypotheses. Hypotheses are a source of *prediction* through deduction. If predictions turn out to correspond with facts time and again, we will ultimately accept the hypothesis. It may then be given the status of a *law of nature*, provided it fits in with a more encompassing body of knowledge, a *theory*. Laws and theories may cover new phenomena we come across. That is, they serve as a basis for *explanation*.

This description presupposes that it is possible to obtain reliable evidence in the form of facts. What facts could have this function? It is tempting to suppose that facts which are open to direct observation are the best source of evidence. However, we should realize that *it is impossible to register facts 'as they really are'*.

Even the most simple observation is loaded with theory (see *2.1*). This is one of the reasons why the above picture of science is a bit naive. It is useful, though, as a preliminary outline. The following example explains it in a more concrete way.

• Example 1. A biologist investigates fluctuations in temperature, oxygen content and other physical factors in the water of rivers. During the investigations she is repeatedly confronted with massive mortality among fish in a particular river. Inspection of her data shows that the oxygen content of the water was always very low during periods with a high fish mortality. There appears to be no correlation between mortality and other physical factors studied. On the basis of these observations the biologist formulates the following hypothesis. 'Low oxygen levels will cause fish to die.'

The hypothesis could be tested by the prediction that continuation of the observations will again reveal a positive correlation between mortality and low oxygen levels. However, this need not prove much since oxygen levels may covary with unknown factors influencing fish. Therefore the biologist decides to test the hypothesis by an experiment. Experiments unlike field observations allow us to keep many factors constant. The biologist predicts that the mortality of fish in aquaria with a low oxygen level will exceed that in aquaria with normal levels. This is indeed what happens in an experiment designed to investigate the hypothesis. The hypothesis is therefore accepted as true by the biologist. Notice that the predictions involve deduction whereas the acceptance of the hypothesis on the basis of confirmed predictions is by inductive reasoning.

Later on the biologist is again confronted with a high mortality among fish in the river. Measurements show that the oxygen content of the water is low. The biologist concludes that her hypothesis explains this episode of mortality. The explanation can be construed as a valid argument. A statement to the effect that low oxygen levels will always lead to mortality in fish, together with the statement that the level is low in a particular case, leads to the conclusion that there must be mortality in this case.

Further research could show how low oxygen levels produce deleterious effects. For example, we could chart specific effects on metabolism. This could result in a theory which unites various generalities. •

I will use the example as a basis for miscellaneous comments on methodology.

1. Logical reconstruction. In the example I neatly ordered various phases of research in a 'logical' way. First we elaborate a hypothesis, then we test it. If it is confirmed we use it as a basis for explanation. Also I explicitly characterized arguments with terms such as 'deduction' and 'induction'. In actual science we will mostly come across a mixture of the phases I distinguished. Moreover, scientists will seldom be fully explicit about their line of reasoning and they will not use much methodological terminology. Therefore scientific work often needs to be reconstructed before we can evaluate it.

For example, suppose the following argument is presented in a scientific paper. 'A massive mortality of fish occurred on July 23 at the site we were investigating. Measurements showed that the mortality was caused by a low oxygen level.' It is probable that this passage must be interpreted as follows. 'A massive mortality of fish occurred on July 23 at the site we were investigating. Measurements showed that oxygen levels were low on this day. The hypothesis that low oxygen levels always cause mortality in fish is well-confirmed. Therefore the hypothesis explains the mortality we observed.' The general hypothesis was omitted in the original formulation. The reader is supposed to read between the lines.

2. Discovery and justification. In the past, philosophers of science used to distinguish two phases of scientific research, the context of discovery and the context of justification. The generation of hypotheses belongs to the first context, the testing of hypotheses to the second one. It is obvious that tests can be evaluated with methodological criteria, for example criteria for the appraisal of predictive arguments. It is much more difficult to evaluate the generation of hypotheses. Hence philosophers claimed that there is a logic of justification but no logic of discovery. Discovery, they argued, is an inherently irrational process.

I would at most defend the weaker claim that in discovery, unlike justification, irrational processes are allowed. Rational inferences, even deductive ones, can be a source of new hypotheses. For the rest, the acceptance of hypotheses which have passed critical tests is also a matter of induction.

In some cases the distinction of discovery and justification does not make sense at all. For example, logical analysis may show that two competing hypotheses which scientists are investigating do not exhaust the possibilities (cf. comments on genetic *versus* environmental determination in *2.5*). Such an analysis, *qua* analysis, would belong to the context of justification. At the same time it would lead to the formulation of new hypotheses and so belong to the context of discovery.

3. Facts. It would be a mistake to regard the 'facts' of science and daily life as things which are 'out there', 'in reality'. Facts are constituted by statements which represent our way of looking at reality. A biologist who notices an object floating at the surface of a river may describe his observation with the statement 'There is a dead fish over there'. A physicist may give a different description: 'The object over there has a specific gravity lower than 1'. The two statements cover very different aspects of the same 'chunk' of reality. Notice that concepts used in either description presuppose theory. The biologist using the concept of fish presupposes a biological classification which he thinks applies to organisms. The physicist's statement is obviously based on the assumption that objects floating on water have a specific gravity lower than one.

4. Generality. The hypothesis that low oxygen levels cause fish to die is a general statement, though my formulation does not reveal this. Its form is better revealed by the following formulation, which nobody will use in ordinary parlance: 'For all

x, if x is a fish, then x will die upon exposure to low oxygen levels'. The generality is due to the presence of a universal quantifier. Hypotheses often have this kind of generality, but they may also have the form of ('general') probabilistic statements (see *4.3*).

Are hypotheses always general in one of these senses? I would say that this is a matter of terminology. If you wish you may call the non-general assumption that somebody has a particular disease a hypothesis. My preference is to use the label of hypothesis for general statements.

Henceforth I will avoid this usage of the term 'general', and use the terms 'universal' and probabilistic' instead, since 'generality' also has other meanings. For example, it is also used for the opposite of specificity. We get a statement which is more general in this sense if we replace 'fish' in the hypothesis I discussed by 'organism'. This statement applies to more entities.

I also have an additional comment on universality. The term 'universal' is often applied to any statement which has a universal component. However, philosophers of science use the term in a more restricted sense which also has my preference. They will only call a statement universal if, in addition, it does not contain an existential quantifier and does not refer to particular times, places or individuals.

5. Inductivism. The foregoing portrayal of science implies that a view which is known as inductivism should be rejected. According to this view, which was inspired by the philosopher Francis Bacon (1561-1626), scientific theories result in a rather straightforward way from an ever-accumulating store of plain facts. Inductivism at its worst vaguely assumes that if we would know 'all the facts', we would thereby have a perfect view of reality.

However, facts are actually loaded with theory right from the start. Expressions such as 'plain facts' and 'all the facts' are hardly meaningful.

6. Objectivity. So far I have disregarded one important label which is commonly used to characterize science, objectivity. 'Objectivity' is by no means an easy concept. It indicates that personal predilections and idiosyncracies are not allowed to interfere with the results of scientific judgement. Subjectivity must be avoided. However, it is difficult to define the boundaries of subjectivity. The stipulation that we are objective when we report facts as they really are, and subjective when we fail to do this, will not do since 'facts as they really are' is a meaningless expression. Facts are loaded with theory. The admonition that, to deal with facts in an objective way, we must not presuppose theories which are inadequate, is not very helpful either. Facts are supposed to be a testing ground for theories, so the admonition will make us run around in circles.

In view of these problems, objectivity nowadays is commonly analysed in terms of intersubjectivity. Our observations are objective if other competent observers will see the same things. Objectivity thus becomes a relative notion. In science you are objective if the facts you report can be recognized by other scientists. As a result, it is not very easy to distinguish science from non-science by an appeal to objectivity.

• *Example 2.* Suppose a 'psychic' tells you that your health must be poor because your aura has the wrong color. Maybe you will not accept this since like many other observers you don't see auras. You may infer that auras don't exist. Most scientists will agree that 'observations' on auras should not be accepted by science. Such observations are not objective in the sense of intersubjective.

But wait a minute. The criterion of objectivity demands that *competent* observers must be able to reach agreement about what they see. Perhaps only psychics are competent in this case. If you let a random sample of people from the street have a look through your microscope, most of them will see the wrong things. They will not be competent since they have not been trained in the use of the microscope, and their knowledge of biology will not be enough to see the right things. There is a difference with psychics, though, because there is no clear-cut, commonly accepted way to become a psychic.

The objectivity of the psychic's observation could be checked by a comparison with other psychics. If it turns out that they see different things the observation will become suspect. •

Exercise

A biologist decides to study life in a particular forest in the hope that this will enable him to formulate new biological theories. Of course he will have to make choices. He could study a random sample of organisms. Alternatively, he could concentrate on birds, or on insects. Other choices to be faced concern the kind of features to be studied, periods of observation, and so forth. Suppose the biologist feels unable to make the right choices and therefore asks your advice. What would your reaction be?

CHAPTER 6

Hypotheses

6.1. Evaluating Hypotheses

Imagine yourself in the following situation. You want to know if a certain substance influences reproduction in freshwater snails. So you expose some snails to the substance and use others as controls. The two groups are kept under carefully controlled conditions (constant temperature, artificial illumination according to a fixed scheme, a strict regime of food and water) and data on reproduction are recorded. There are differences as you expected. You are also surprised by a phenomenon you did not expect. There are clear non-random fluctuations in reproduction in the two groups. Each snail produces a couple of egg masses in each week, but the differences between weeks are quite large. Yet the conditions are as constant as you could get them. The phenomenon intrigues you, and you decide to investigate it. How to proceed? First of all you will need a hypothesis which could account for the data. How does one get at a good hypothesis? Would you have a suggestion?

Well, I hadn't when I was in this situation back in the 1960s when I did graduate research in biology. Fortunately, one day my professor was struck by an idea when we were thinking about the fluctuations in reproduction. He challenged me to mention one factor I definitely had not kept constant in the experiment, and which might influence reproduction. I should have been able to mention many factors, but not a single one came to my mind. "Here's what you have been overlooking", the professor said, "atmospheric air pressure". I certainly would not have thought of that.

I managed to lay my hands on records of air pressure and, sure enough, it turned out that the fluctuations in reproduction were correlated with air pressure. Not with the pressure height, though. There was a significant correlation with the amount of *change* in atmospheric air pressure. Now one experiment does not say much, so I did several additional experiments, which confirmed the correlation (for details see Van der Steen, 1967).

Afterwards it struck me that this is a weird phenomenon. The snails continually move up and down through the water in their jars. This exposes them to pressure changes which far exceed changes in atmospheric air pressure. It is unclear therefore how relatively minor fluctuations in air pressure could affect the snails. If I would have thought of this earlier, perhaps I would not have bothered about additional experiments. After all, the correlation initially observed could have been due to chance.

However, I now had a phenomenon on my hands which cried out for an explanation. I surmised that some unknown factor correlated with pressure change had to be responsible for the effects on reproduction. After an extensive study of the literature I found out that pressure changes are correlated with changes in subtle geophysical factors (cosmic rays, electric fields, magnetic fields). Also, there appeared to be many data showing that such factors do influence organisms. I had not known these things beforehand. All in all it is probable that subtle geophysical factors affect reproduction in freshwater snails. I did not further investigate this possibility for lack of appropriate equipment. As far as I know the issue is still open.

If I would have known the literature before doing the additional experiments, I could have inferred from it that reproduction may well be correlated with changes in air pressure. Thus the professor's lucky guess could have been arrived at in a quite different way.

This little story is representative for scientific research in that it shows that processes of discovery are often capricious. Thus there are no strict rules for the generation of hypotheses. Uncontrolled flashes of insight, a thorough study of the literature, inferences from data, and many other things may yield hypotheses worthy of further study.

Once we have a hypothesis, however, there are strict *methodological criteria* for dealing with it. Most important is the requirement that hypotheses must be subjected to tests in a rigorous ways. That is possible only if they satisfy the criterion of *testability*. Testability and the process of testing are at center stage in this chapter. Hypotheses are not properly testable unless various other methodological criteria are satisfied. I will discuss these criteria first.

Perhaps it will strike you as odd that generality is missing in the list of criteria presented below. I am *presupposing* here that hypotheses satisfy this criterion since I am using the term 'hypothesis' for general statements. The subject of generality is discussed extensively in the next chapter, where it is shown that 'generality' is a composite notion.

First and foremost, a hypothesis will have to satisfy the criterion of clarity. The criterion applies to the logical structure of hypotheses and to the concepts they contain.

The example from ecology below illustrates the importance of clear concepts. Concerning clarity, awareness of the logical structure of hypotheses is at least as important. As I argued in *4.2* the structure of hypotheses may be more complex than formulations in ordinary language suggest. Thus it is easy to overlook the fact that the statement 'All features of organisms have a function' contains an existential quantifier ('there is ...') besides a universal one ('all ...'). This becomes clear upon reformulation: 'For *all* features of organisms *there is* a function'. Later on I will show that quantifiers have important consequences for testability.

• *Example 1.* From the fifties onwards, many ecologists have defended the thesis that complex communities are relatively stable: complexity begets stability. However, consensus about the thesis was never reached. Time and again, opponents would uncover evidence allegedly discrediting it. In recent years, it has been realized that conceptual issues were at the bottom of the controversy. 'Complexity' and 'stability' are ambiguous concepts, they are used in many ways. Hence the thesis is ambiguous, and different researchers had been addressing different questions.

A lucid overview is given by Pimm (1991). He suggests that ecologists have used the word 'stability' to mean at least five different things. First, there is stability in the mathematical sense, which implies that variables return to equilibrium conditions after displacement from them. Second, 'stability' can stand for 'resilience', defined as how fast a variable that has been displaced from equilibrium returns to it. Third, 'stability' in the sense of 'persistence' indicates how long a variable lasts before it is changed to a new value. Fourth, there is 'resistance', which stands for the impact of changes in a variable on other variables. Fifth, 'variability' is the degree to which a variable varies over time.

To complicate matters, studies concerning the relations between complexity and stability have concerned three levels of ecological organization. Also, 'complexity' has been taken variously to mean the species number in a system, the connectance of the food web, and the relative abundances of species in a community.

> In short, ... ecologists were looking at the combinations of five definitions of stability, three definitions of complexity, and three levels of organization—a total of forty-five possible questions about the relationships between community complexity and stability. Rarely did two ecologists look at the same question, although when empirical and theoretical ecologists *did* look at the same relationships between complexity and stability there was remarkably good agreement (Pimm, 1991, p. 15). •

I want to make some remarks on context-dependence before I introduce additional methodological criteria. In chapter 2 I have argued that *the importance of clarity is context-dependent.* Hence the criterion of clarity for hypotheses should not be interpreted in an absolutist way. Considering the concepts in the example we should insist that the criterion of clarity is of paramount importance. The decision to make it operative in this case may imply that other criteria get a lower priority; seldom if ever will it be possible for scientific work to maximally satisfy all methodological criteria at the same time. I have explicitly dealt with this aspect of context-dependence in various chapters (for example chapters *1* and *7*). Here I will not do that because it would make the text excessively complicated.

Next, hypotheses must have empirical content. In other words, hypotheses must not be logical statements. According to this criterion their truth or falsity should not depend on meanings or logical form alone. Example 2, admittedly a controversial one, illustrates this for the concept of fitness, which plays a crucial role in evolutionary biology. Example 3, which also concerns evolution, is a more perspicuous illustration.

• *Example 2.* The theory of evolution is often attacked on the ground that it is based on the hypothesis that 'the fittest survive', a statement which is regarded as a logical statement (a tautology) by many authors. Thus it has been argued that 'survival' (a shorthand for 'reproductive survival', that is 'number of descendants') occurs in the definiens of 'fitness'.

The following definition does capture the way the concept is often used in population genetics. 'The fitness of organism x is greater than the fitness of organism y (x and y being members of the same species) in environment E' $=_{df}$ 'x has a greater expected number of descendants than y in E'. Notice that the term 'expected' is essential. If one member of an identical twin has no descendants because she is struck by lightning, whereas her sister is blessed with progeny, we will not attribute a lower fitness to her. The assumption that organisms with the same features could have different fitnesses is counter-intuitive.

Many authors have tried to rebut the tautology charge. As far as I know all of them have concentrated on the concept of fitness in the thesis that the fittest survive. Many alternative fitness concepts have been proposed (for some examples, see Byerly and Michod, 1991). It is unclear whether any of them would make the charge harmless; juries in biology and the philosophy of biology are still out.

The discussions keep surprising me since the disputants take the concept of survival for granted. It seems to me that the solution of the tautology problem is easy once we recognize that the term 'survival' is ambiguous. First, there is survival in the sense of having descendants, the notion that generates the problem. Second, there is the survival of types in populations, which is a very different notion. Survival in the latter sense is determined by survival in the first sense *and* factors such as migration and mutation. Thus 'The fittest survive', on this interpretation, becomes a false empirical statement unless qualifications are added!

Population geneticists and evolutionary biologists will use fitness data to explain survival in the second sense. This is perfectly legitimate. 'The fittest survive' thus can be construed as a plain empirical statement, one which happens to be false unless one adds a clause to the effect that there is no migration, mutation etc.

In view of the pervasiveness of controversies concerning fitness I do not know whether my solution of the tautology problem would be accepted by the parties involved. However, my comments suggest that analyses focusing on the concept of fitness alone are problematic. Tautology charges remain unfounded if the concept of survival is taken for granted. •

• *Example 3*. The fossil record suggests that the diversity (number of species) of higher taxa (families, orders and the like) tends to follow a sigmoid pattern in the course of time. After a slow start there is a rapid increase in species numbers which subsequently levels off and tends to a limit in the end. Would the hypothesis that the diversity of higher taxa follows a sigmoid trend be acceptable?

The graphs which have been produced suggest that the issue is clear-cut, but appearances are deceptive. It is conceivable that diversity is a criterion for taxonomists to construct classifications in particular ways. Thus they could decide to split taxa with a very high diversity. This could explain the fact that diversity tends toward a limit. If this would be so, the hypothesis would not be a purely empirical one. •

According to the criterion of *simplicity*, which is illustrated in the next example, we should prefer a hypothesis which is simple in form over a more complex one if other things are equal.

• *Example 4*. Suppose we are interested in the relation between temperature in a 'normal' range and locomotory activity in a particular beetle. An experiment shows that data obtained at four temperatures are compatible with the hypothesis that activity increases linearly with increasing temperature. That is, in a graphical presentation all the data are close to a straight line and statistical analysis shows that deviations are not 'significant'.

From a mathematical point of view such data can also be represented by an infinite number of more complex functions. However, we will stick to the linear function because it is the simplest one. There is nothing in the data which compels us to do this. The decision is motivated by the desire to keep scientific hypotheses and theories as simple as possible. •

We should realize that simplicity of mathematical form is not the only species of simplicity. Simplicity may also relate to numbers of factors postulated in a hypothesis, to the form of definitions of concepts in a hypothesis, and so forth.

Hypotheses will be accepted only after they have survived critical tests. This implies that they must be testable in the first place. Testability and testing will be discussed in the next section.

If a hypothesis is accepted we will mostly want to accommodate it in a theory. For that purpose we will have to consider additional criteria such as generality, coherence, explanatory power and predictive power. These are discussed in later chapters.

Exercise

6.1. Evaluate the following hypotheses with respect to logical form. Are all the hypotheses acceptable?

    a. Kidneys remove toxic substances from the blood.
    b. Every prey species has at least one predator.
    c. Life is present on other planets than the earth.
    d. All species show reproduction.
    e. Organisms get ill after the consumption of toxic substances.
    f. Plants which perform photosynthesis contain starch.

6.2. Testability and The Process of Testing

On a few occasions I have found beautiful mushrooms growing in washrooms; I mean, really growing *in situ*, from wooden fixture. Many hypotheses could account for this phenomenon. Let me consider one candidate. If the humidity in any washroom is excessively high, then mushrooms grow on the wooden fixture if there is any. This hypothesis is somewhat imprecise of course, but I let that pass. We could make it more precise by specifying criteria for excessive humidity and other elements mentioned in the hypothesis.

Here is one way in which we can test the hypothesis. We set out to visit as many washrooms with wooden fixture in them as we can. We predict that all washrooms we visit that have an excessively high humidity, have mushrooms growing in them. The prediction will be regarded as false if we find any washrooms with an excessively high humidity but no mushrooms on the wooden fixture. If it turns out to be false, we can conclude that the hypothesis must also be false. In that case we will have to replace it by an improved version or a wholly different hypothesis, which in its turn can be subjected to tests.

The mushroom case as presented charts the bare bones of hypothesis testing and, if elaborated somewhat, it will indicate why bare bones don't suffice. I suggest that you keep it in mind and use it to implement for yourself the relatively abstract schemes I will present. I will supplement it with other, more realistic examples to explain the schemes. Let us begin with the bare bones.

Hypotheses are tested by comparing predictions based on them with facts. The predictions are called test implications.

In the most simple situation, the process of testing is as follows. From a hypothesis H we infer the test implication I. If it turns out that I is false, we conclude that

H must be false. The argument leading to this conclusion has the form of the modus tollens:

$$\frac{\text{if } H \text{ then } I}{\text{not-}I}$$
$$\overline{\text{not-}H}$$

According to this scheme hypotheses are apparently testable in the sense of *falsifiable*. It is possible to show that a hypothesis is false *if* it is false. The test implication is said to falsify the hypothesis. If *I* turns out to be true we should not conclude that *H* is true. That would amount to a fallacy of affirming the consequent. If *H* is a universal statement it is not testable in the sense of *verifiable*, because evidence bearing on it would be inexhaustible. It would be *confirmable* though, that is, we could obtain evidence in its favor.

• *Example 1*. Concerning the metabolism of some species of animal we hypothesize that a particular chemical compound *A* is the direct product of another compound *B*. To check this we inject animals with an amount of radioactively labeled *B*. A test implication of the hypothesis is that *A* will be the first substance which will also show radioactivity. *A* is indeed the first radioactive substance we are able to isolate. This confirms the hypothesis.

A full-fledged reconstruction of this example, which I will not present here, would show that it is much more complex than the scheme presented above. All the same it is clear that the example involves confirmation. •

Awareness of logical form is indispensable if we want to understand the process of testing. If the statement 'All swans are white' would be true (it isn't), then we would not be able to prove it. It is not verifiable because it is universal. However, if it is false (it is) we can prove that in principle through the observation of a specimen that isn't white. Because it is universal it is falsifiable. Conversely, existential statements such as 'There is a black swan' are verifiable, not falsifiable. The swan case is discussed in a more technical way in the example below.

• *Example 2*. From the universal statement that all swans are white we can infer test implications such as 'If *a* is a swan, then *a* is white'. In symbols: the statement '$(x)(\text{if } Sx \text{ then } Wx)$' has implications such as 'if Sa then Wa', 'if Sb then Wb', ... The universal statement is falsified if we observe a black swan. If we would only observe white swans that would confirm the statement. Verification would be impossible since it remains possible that we will come across a non-white swan in the future. (There are black swans as a matter of fact.)

In brief, universal statements such as the one considered here are falsifiable, not verifiable. We can show that they are false *if* they are false, but it is impossible to show that they are true *if* they are true. Existential statements, in contrast to this, are verifiable but not falsifiable. The statement that there is at least one black swan, '$(\exists x)(Sx \text{ and } Bx)$', is verified by the observation of a single black swan. If it would be false we could not falsify it, for it would remain conceivable that we will observe a black swan in the future. •

The swan case is simple.

In science we seldom come across simple situations . Strict verification and strict falsification are borderline cases. We normally should be content with confirmation, or disconfirmation as the case may be. Accordingly we will accept or reject hypotheses after confirming and disconfirming evidence have been weighed.

I will elaborate this in the sequel, but I want to concentrate first on some details concerning terminology and logical form.

The concepts of verifiability, falsifiability and confirmability stand for types of testability. We should not confuse them with the concepts of verification, falsification and confirmation, which denote the results of actual tests.

A hypothesis is testable if it is possible to test it. The term 'possible' stands for *'logically possible'*. The demand that tests be *empirically possible* will normally be regarded as unduly strong. As the foregoing example shows, an inspection of logical form indicates what logical possibilities there are for tests. Empirical possibilities are more elusive. A few decades ago tests of the hypothesis that there is life on other planets than the earth were empirically impossible. By now such tests are feasible, at the very least for planets in our solar system.

The presentation of relationships between logical form and various kinds of testability in example 2 is a bit simplistic.

First, I have assumed that we are dealing with statements that do not refer to particular places, times or individuals. A statement to the effect that there is a match in a *particular* box, though existential, is easily falsified.[*]

Second, I have only considered affirmative statements. Notice that the universal statement that all swans are white is equivalent to the existential statement that there are no non-white swans. Hence negative existential statements, unlike positive ones, are falsifiable, not verifiable. And so forth.

The following table gives a more accurate picture of the impact of logical form on falsifiability and verifiability. To get the connection with the swan example you should replace 'Px' in the table by 'if Sx then Wx'.

---

[*] I owe this point to Cor Zonneveld.

		verifiable?	*falsifiable?*
a1 $(x)(Px)$	a2 $(x)(\text{not-}Px)$	no	yes
b1 $\text{not-}(\exists x)(\text{not-}Px)$	b2 $\text{not-}(\exists x)(Px)$	no	yes
c1 $(\exists x)(Px)$	c2 $(\exists x)(\text{not-}Px)$	yes	no
d1 $\text{not-}(x)(\text{not-}Px)$	d2 $\text{not-}(x)(Px)$	yes	no

The following expressions in the table are logically equivalent: a1 and b1; a2 and b2; c1 and d1; c2 and d2 (see *4.2*).

Back to the main issue. How should we think about the criterion of testability? Some decades ago philosophers of science used to interpret the criterion of testability as a criterion of verifiability. On this interpretation the criterion is definitely too strong since hypotheses tested in science are typically universal or probabilistic statements. Neither kind of statement is verifiable. The philosopher Popper (1959, 1963) therefore proposed that verifiability be replaced by falsifiability as a criterion. However, as I will argue below, this criterion is also too strong.

Testability must be construed as a relatively weak criterion of (dis)confirmability: a hypothesis is inadequate if a reconstruction of its form shows that it is logically impossible to find evidence for or against it. It seems to me that this criterion will be automatically satisfied by hypotheses which are clear and have empirical content.

Many scientists still endorse a stronger testability criterion. Specifically, in the life sciences many researchers subscribe to (a simplistic version of) Popper's falsifiability criterion. In view of this some objections against the criterion are discussed here to get a better perspective on testability in science.

First objection against the criterion of falsifiability

In the practice of science, many assumptions are needed in the derivation of test implications from hypotheses. The mushroom story at the beginning of this section is a case in point. As I said I won't elaborate it. Maybe you want to do so for yourself.

Therefore the simplistic scheme at the beginning of this section must be replaced by the following one.

$$\frac{\text{if } H \text{ and } A, \text{ then } I}{\text{not-}H \text{ or not-}A}$$

'*A*' in the scheme stands for assumptions that must be made. If a test implication turns out to be false, we are allowed to conclude only that the hypothesis *or* an assumption is false. Therefore the hypothesis can be falsified only if the assumptions can be verified. That is impossible in many cases. For example, in experimental work we will have to assume that the equipment we use is properly working, and that unknown factors are not interfering with the results. In brief, one of the assumptions is that *there are* no disturbing factors. We are dealing here with an unverifiable existential statement.

The improved scheme does not cover all cases of hypothesis testing. Hypotheses can also be (dis)confirmed by theories. (Recall the mushroom story!) All in all we will need fairly complex schemes if we want to account for any particular case of hypothesis testing in a realistic way. I do not want to complicate my exposition any further, so I will not present schemes accounting for the role of theories. The following example makes the scheme presented above more concrete.

• *Example 3.* In the past I have done research designed to show that population density affects growth in snails. To investigate this I compared two groups of the freshwater snail *Lymnaea stagnalis* at a density of two per jar and ten per jar, respectively. The experiment was set up in such a way that conditions were similar in other respects—to the extent one can realize this. Thus the mean food consumption was kept constant. Amounts of food in the high density jars were five times the amounts in low density jars, and the amounts were such that all the food was eaten. I assumed (!) that differences in amounts of food present do not in themselves influence growth.

My hypothesis was that high density will repress growth. Indeed it turned out that growth was much lower in the high density jars than in the low density jars. This seems to confirm the hypothesis.

In reality the results need not be due to density at all. It is possible to show by statistical reasoning that the *variability* in food consumption will be different at different densities, even if density does not causally affect individual snails. A mathematical model showed that this 'sampling artefact' may result in differences in growth, if the relation between food consumption and growth is nonlinear (Jansen and Van der Steen, 1975). So density need not have an effect after all since there is a disturbing factor in the form of an artefact.

The example shows how difficult it is to realize conditions in which effects of disturbing factors are excluded. •

As argued above, testability must be construed as a rather weak criterion, so that hypotheses which are clear and have empirical content are *ipso facto* testable. This does not imply that there are no problems with testability in practice. Apart from troubles with practical feasibility, there is the problem that *it is possible to make a hypothesis untestable by manipulating assumptions. An investigator who cherishes a particular hypothesis may assume that adverse evidence he comes across is due to, say, shortcomings of the equipment he is working with.*

Assumptions which are invented in the face of negative evidence are called *ad hoc assumptions*. Researchers should try to minimize the role of such assumptions. Examples 4 and 5 illustrate this.

• *Example 4*. Dr. Johnsen, a general practitioner, elaborated the hypothesis that diseases of internal organs are associated with pains in particular fingers. Specifically he thought that malfunctions of the heart lead to pain in the little finger of the left hand.

The theory is an example of 'projection theories' which are entertained in some circles in alternative medicine. Thus, according to some researchers, the body is projected onto the foot sole and onto the ear.

To test the theory, Johnsen gives a number of patients a routine check-up. He asks them to spread their fingers on a table. The fingers are gently struck with a little hammer to detect abnormal sensitivities. In this way Johnsen diagnoses malfunction of the heart in many patients. The afflicted persons are sent to a cardiologist. After a while the cardiologist gets fed up with this. "What do you think you are doing?", he says to Johnsen, "these people are all in good health". Johnsen replies that his method is obviously very sensitive, since it permits one to diagnose heart disease in an early stage when orthodox methods fail.

Johnsen's hypothesis is an empirical statement which is clearly testable. However, he treats it as if it were a logical statement since he regards pain in a particular finger as an overriding criterion of heart failure. Thus the hypothesis is made untestable in practice. If he is confronted with negative evidence, Johnsen saves the hypothesis by the *ad hoc* assumption that orthodox methods of diagnosis fail to reveal all cases of heart malfunction. Needless to say, this procedure is not acceptable. •

• *Example 5*. 'Morphological changes which occur in the course of evolution are always gradual rather than abrupt.' Many evolutionary biologists would accept this hypothesis. Let us assume that it is sufficiently clear.

To test the hypothesis, we will need to consider data from the fossil record. The record is notoriously incomplete, so we will come across many apparent discontinuities. Therefore the hypothesis can always be protected against negative evidence, by the *ad hoc* assumption that transitional forms happened to leave no fossils. In the absence of additional arguments that would be a poor basis for accepting the hypothesis.

Fortunately there are additional arguments. First, one can determine the age of strata in ways that do not depend on the hypothesized age of fossils. In some cases the absence of transitional forms in some location appears to be connected with the absence of strata with a particular age. Second, according to a theory which has much evidence in its favor, new species often originate in isolated areas. They may spread to other areas after an appreciable amount of morphological change has taken place. This theory therefore predicts that one *should* find breaks in the fossil record in some places, even if morphological change is gradual.

In spite of all this there are controversies over the hypothesis (see *2.4*, exercise 2; *6.3*, example). Some researchers think that rapid changes during speciation events are alternated with periods of stasis. This is a reason for them to reject the hypothesis. However, we should realize that apparent abruptness of morphological change during speciation is projected onto a geological time-scale. It is surely compatible with gradual change on a 'normal' time-scale. •

Second objection against the criterion of falsifiability

Hypotheses often have a logical form which precludes falsification. Probabilistic hypotheses are an example. In addition to this, we cannot falsify a hypothesis which contains an existential quantifier beside a universal one. An example is given below.

• *Example 6.* The hypothesis that all features of organisms have a function, which I discussed in *4.2*, has an existential besides a universal component ('for all features, *there is* a function ...'). Therefore it cannot be falsified. Some researchers would regard it as unacceptable *because* it is unfalsifiable. I would not agree because the falsifiability criterion is unacceptable for other reasons.

Anyway it is possible to find evidence for or against the hypothesis. If we manage to attribute a function to many features of organisms this will confirm the hypothesis. If, *despite strenuous efforts*, we would not uncover a function in many cases, we would have good reasons to reject the hypothesis.

The presence of an existential component makes *ad hoc* reasoning easy. We can simply *assume* that seemingly functionless features have an unknown function. That would constitute a poor basis for accepting the hypothesis. It is not necessary, though, to succumb to this line of reasoning. •

Third objection against the criterion of falsifiability

The essential role of theoretical concepts in science makes falsification impossible. Hypotheses often contain theoretical concepts which are connected with the level of observations in indirect and complex ways. Moreover, concepts used to describe observations are loaded with theory (see *2.1*). The example below illustrates this.

• *Example 7.* Consider the hypothesis that hunger decreases the secretion of a particular hormone in some species. One way to test such a hypothesis is by a histological investigation of tissue responsible for the production of the hormone. The test implication will be that secretory activity is lower in cells of starved animals than in cells of animals receiving a normal diet. This implication is itself a hypothesis which we may test by studying, say, the density of granules of a particular kind in histological preparations. Such a test would be based on the assumption that the granules represent secretory material. The assumption again represents a hypothesis, one that will have been tested in different experiments. •

The example shows that a test implication of a hypothesis can itself be a hypothesis; hypotheses can form hierarchies. Moreover, we may need additional hypotheses as assumptions for the derivation of test implications from a hypothesis under investigation. The example also shows that the complexity of the testing process is associated with the presence of theoretical concepts in hypotheses.

In the philosophy of science the objections against the criterion of falsifiability have been known for a long time (see e.g. Harding, 1976). Most philosophers accept them. In spite of this the criterion kept playing a role in the life sciences. Many biologists have explicitly endorsed it in articles in *Systematic Zoology* in the seventies and the eighties. Some leading epidemiologists have likewise defended it in these years. In various other areas of biology and medical science testability is seldom discussed.

Exercises

6.2.1. Prepare a list of assumptions that play a role in example *1*.

6.2.2. Evaluate the following test of the hypothesis that the fly *Musca domestica* is able to perform orientation with the help of magnetic fields.

Three test implications are derived.

    a. '*Musca domestica* has a sensory organ that registers magnetic fields. Electrophysiological recordings will show that signals from one of the sensory organs will change when flies are put into a magnetic field.' All the known sensory organs are investigated. No change in signals is found. It is concluded that the test implication is false.

    b. 'If individuals of *Musca domestica* are put into a magnetic field, the number of flies with the body axis aligned to the direction of the field will significantly deviate from the number expected under a random distribution.' This implication appears to be true.

    c. 'The nervous system of dead *Musca domestica* is inactive. Orientation without an active nervous system is impossible. Hence dead flies should not orient toward a magnetic field.' Experiments show that dead flies do orient in this way. (If you are surprised, read Becker and Speck, 1964, who observed and explained this phenomenon.)

The hypothesis is rejected since two out of three test implications are false.

6.3. Alternative Hypotheses

In actual research we will seldom be dealing with a single hypothesis. Ideally there are alternative competing hypotheses which form an exclusive and exhaustive classification. Clusters of hypotheses which do not constitute an exhaustive classification may lead to bias.

In a statistical evaluation of hypotheses researchers commonly work with a dichotomous classification consisting of a *null hypothesis* and an *alternative hypothesis*. I will not deal with statistics in detail; many excellent texts on the subject are available. The following example suffices by way of a general outline.

Suppose we want to check whether a particular coin is biased, that is, whether the probabilities of throwing head and throwing tail are unequal. The null hypothesis in this case will be that $p(\text{head}) = 0.5$, that is, that the outcome will be head in 50% of an infinite number of throws. Rejection of the hypothesis would imply acceptance of the alternative hypothesis that $p(\text{head}) \neq 0.5$.

The hypothesis can be tested by considering a sample of, say, ten throws. To evaluate it we need a *test statistic*, a variable that takes on a certain value depending on the outcome of the test. If we assume the null hypothesis to be true, we may compute the probabilities of the various values of the test statistic. Thus we can differentiate between probable and improbable outcomes. In the present case, the number of heads in ten throws is a suitable test statistic. We can compute probabilities for all possible values of this statistic under the null hypothesis that $p(\text{head}) = 0.5$. These values jointly form a *probability distribution*. The alternative hypothesis is actually a composite one (*p* can take on many values), so it is not associated with one particular probability distribution. That is precisely the reason why it is tested *via* an evaluation of the null hypothesis.

If the outcome of the test belongs to a set of extreme outcomes which are improbable under the null hypothesis (outcomes in one of the tails of the probability distribution), the null hypothesis will be rejected. The set of extreme outcomes is mostly chosen in such a way that it comprises 5% of all possible outcomes. This means that the probability that the null hypothesis is rejected for the wrong reason is 5%.

In standard statistics, a null hypothesis and its alternative constitute a neat dichotomous classification which is exclusive and exhaustive.

Statistical hypotheses mostly function as test implications of more general, abstract hypotheses. Researchers considering these more general hypotheses do not always deal with alternatives in a proper way. At times the alternatives pitted against each other do not exhaust the possibilities.

In technical terms, such alternatives are *contraries* rather than *contradictories*. The statements 'A is white' and 'A is black' are contraries; 'A is white' and 'A is nonwhite' are contradictories. The difference between contraries and contradictories is that two contraries can *both* be false, whereas in the case of contradictories one of the alternatives must be true. A feature shared by contraries and contradictories is that the alternatives considered cannot both be true.

Exhaustiveness and other features discussed above are *logical* criteria. We should realize that a classification of hypotheses which does not exhaust the possibilities from a logical point of view, may still be adequate because it is exhaustive from an *empirical* point of view. There may be good grounds for researchers to test a nonexhaustive array of hypotheses, if the alternatives disregarded are implausible for empirical reasons.

I will show by one longish example from the literature that the subject of alternative hypotheses is important in the evaluation of research.

• *Example*. In *2.4*, exercise 2, I introduced two alternative hypotheses concerning speciation which were pitted against each other by Eldredge and Gould (1972). In their formulation the hypotheses are as follows.

Phyletic gradualism
1. New species arise by the transformation of an ancestral population into its modified descendants.
2. The transformation is even and slow.
3. The transformation involves large numbers, usually the entire ancestral population.
4. The transformation occurs over all or a large part of the ancestral species' geographic range. (...)
Punctuated equilibria
1. New species arise by the splitting of lineages.
2. New species develop rapidly.
3. A small sub-population of the ancestral form gives rise to the new species.
4. The new species originates in a very small part of the ancestral species' geographical extent—in an isolated area at the periphery of the range.

The hypotheses need clarification. For example, splitting of lineages is not mentioned under phyletic gradualism. However, we can hardly assume that gradualism denies that any splitting or some such process does occur. Denying this would imply that, if life originated once, we should still be left with a single species!

For simplicity of presentation I will assume that the four statements under punctualism are meant as denials of the four statements under gradualism.

Eldredge and Gould's view was attacked soon after being published and it remains a source of controversy. Here I will consider the criticism mounted by Harper (1975) in the leading journal *Science*. (I have slightly adapted Harper's formal notation so that it fits conventions adopted in this book.)

According to Harper, Eldredge and Gould's presentation is misleading because the two hypotheses are contraries rather than contradictories. He argues his case with the help of the following formalism. Punctualism is represented by the expression $p(S_i/S) = 1$. 'S' stands for the set of sexually reproducing Metazoa, 'S_i' for the subset which originated in accordance with punctualism. The expression $p(A/B)$ is used for the conditional probability that A, given B. (Harper allows for the possibility that, according to punctualism, the p-value may slightly deviate from 1; this is immaterial in the present context.) Gradualism, according to Harper, is represented by the expression $p(S_i/S) = 0$ or, equivalently, $p(\text{not-}S_i/S) = 1$.

These formulations show, Harper says, that the two hypotheses are contraries rather than contradictories, for the simple reason that $p(S_i/S)$ could take on any value in the range from 0 to 1. Punctualism and gradualism thus represent extremes of a spectrum of possibilities.

Harper's conclusion that we are dealing with contraries is obviously right. However, there is a strange element in his line of reasoning. The use of 'S_i' for one model and 'not-S_i' for the alternative implies that the statement that a *particular* species originated in the punctualist way, and the statement that it originated in the gradualist way, are contradictories rather than contraries. However, in reality alternatives concerning individual species are contraries. Harper should have used the symbol 'S_j' instead of 'not-S_i' because there are more than two logical possibilities for speciation. In short, Harper himself makes the mistake he is uncovering.

Eldredge and Gould's hypotheses are composite ones consisting of four different statements. Gradualism can be represented by the following scheme:
'$(x)(Px$ and Qx and Rx and $Sx)$', 'x' being a variable for species. Punctualism, analogously, is covered by '$(x)(\text{not-}Px$ and not-Qx and not-Rx and not-$Sx)$'.

Harper only pays attention to the first part (the universal quantifier) of these statements. His view amounts to the thesis that there are other hypotheses which do not contain a universal quantifier. He implicitly assumes, however, that concerning the second part of the statements there are only two possibilities: 'Px and Qx and Rx and Sx' and 'not-Px and not-Qx and not-Rx and not-Sx'. Thus he takes 'not-p and not-q and not-r and not-s' to be the negation of 'p and q and r and s'. However, the proper negation is 'not-p or not-q or not-r or not-s'.

The analysis shows how logic can help us unmask false dichotomies. At the present time researchers are still pitting gradualism and punctualism against each other. It is obviously difficult to get rid of bias associated with dichotomies. •

Non-exhaustive arrays of alternative hypotheses and theories are common in science. Likewise for explanations (see *8.2*). This is indeed an innocuous feature of science as long as we are aware of its prevalence. When it is overlooked, the chances are that our research will be pervaded by bias.

Exercise

6.3. Which of the following pairs of hypotheses constitute proper alternatives?

  a. Schizophrenia is genetically determined *versus* schizophrenia is environmentally determined.
  b. Geographical isolation is necessary for speciation *versus* geographical isolation is not necessary for speciation.
  c. Foraging behavior is altruistic *versus* foraging behavior is egoistic.
  d. Encephalitis is caused by a viral infection *versus* encephalitis is caused by a bacterial infection.

6.4. Controlled Experimentation

The experimental manipulation of factors that might influence a phenomenon we are interested in is one of the most powerful tools to test hypotheses. However, experimental tests, no less than other tests, are sometimes beset with pitfalls.

To illustrate this I will discuss tests of hypotheses that particular medical treatments, say with drugs, are effective. As the example below shows, effects observed after a treatment are not hard evidence in themselves.

• *Example 1.* Suppose you visit a doctor because you have a persistent headache. The doctor prescribes pills and behold, after you have taken a few of them, the headache disappears. Would it be proper to conclude that the pills work? Of course not. It is possible that the headache would have disappeared anyway. Also, there may be an effect of the affection bestowed on you by the doctor. •

We should be cautious with the interpretation of effects observed after a treatment, especially when there is no background theory explaining treatment effects. (Remember that theories can play an important role in the testing of hypotheses!)

The best way to investigate effects of a treatment, say with drugs, is by comparing patients who receive the treatment with untreated controls. If we compare a treatment group with a no-treatment group, we should ensure that the groups do not systematically differ in other respects. For example, we should not provide women with a treatment and use men as controls. The best set-up is one in which patients are randomly assigned to groups.

However, this will not suffice to exclude all unwanted systematic differences. The mere fact that patients know whether or not they are getting a drug may have effects. So the patients should be ignorant. This can be achieved by giving patients in the control group, without their knowing it, a pseudo-treatment in the form of 'drugs' without active substances. Such a treatment is called a *placebo*.

Even this set-up is not air-tight. If the physician knows what he is prescribing, he may unwittingly deal with patients in the two groups in different ways. This may obscure results. Moreover, he may tend to interpret the results of treatment and pseudo-treatment in a biased way. Hence the physician, like the patients, should not know whether he is prescribing a drug or a pseudo-drug. Experiments in which all this is realized are called *double-blind experiments* since both parties are 'blind'.

Unfortunately even double blind-experiments are not air-tight. Consider the following example.

• *Example 2.* Suppose you have a headache again. The doctor you consult has a request. She is investigating a new drug and she wants you to participate in a double-blind experiment to test its effectiveness. You agree, and the doctor provides you with pills, nature unknown. The day after the beginning of the (pseudo-?)treatment you feel tired. Your first thought is that you are experiencing side-effects of the pills you are taking. So you assume that you have been provided with the real thing. You feel relieved, and the headache is gone after a while. Maybe the drug is responsible for this. However, don't underestimate the benefit of feeling relieved. •

Within medicine, double-blind experimentation is widely regarded as the paradigm of sophisticated scientific work. Sophisticated it is, but there are limitations. An *extensive methodological literature, mostly disregarded in medicine*, shows that we must be very cautious in interpreting results of double-blind experiments (see e.g. White, Tursky and Schwartz, 1985). Thus the mere fact that patients know that they are in an experiment may have profound effects.

It is possible to take the knowledge factor into account by experimental set-ups which are even more complex. However, the pervasive application of ever more sophisticated designs would soon make medical research unmanageable. In fact a set-up without limitations is impossible. We will always need to make choices concerning variables deemed relevant, time-scales, categories of patients to be treated, and so forth.

The example below is a case in point.

• *Example 3.* It has been shown that the knowledge factor does have marked effects. Experiments on effects of neuroleptics, a class of psychopharmaceuticals, on schizophrenia are an example. In a review of results, Paul (1985, p. 145) notes that different forms of controlled experimentation have led to very different results. Almost all possible outcomes, from quite adverse to quite positive and anything in between, have been observed at one time or another. •

Experimental work on medical treatments is meant to improve on non-experimental work which relies on an intuitive, commonsensical interpretation of anecdotal evidence. It is generally thought that intuition is unreliable. I would argue that *we need both experimental and anecdotal evidence. Each has its strong and its weak points.* Anecdotal evidence concerning the response of patients to treatments may be relatively rich, but its force is weak. Experimental evidence has more force, but it is less rich since we can only consider a limited number of variables.

I will illustrate strong and weak points of intuition and common sense by two examples.

• *Example 4.* Experimental psychologists have documented shortcomings of human judgement in many ways. Physicians have not been spared by them. One example is extensively discussed in a book by Faust (1984). He describes an experiment in which two groups of specialists in internal medicine were asked to diagnose complaints of patients with known diseases. The specialists were not informed about the nature of the diseases. One group was allowed to use any data that might assist in diagnosis, for example results of many different blood tests. The other group only received general information concerning the background, for example socioeconomic status, of patients and complaints felt by them. No significant difference in the percentage of correct diagnoses was found between the two groups. Mistakes were common in either case. According to Faust the explanation is that humans are unable to consciously process complex information in an unbiased way.

Faust's analysis does not imply that the processing of complex information will *always* make judgements unreliable. Subconscious processing may be quite reliable indeed. Field biologists are often able correctly to identify species they casually observe. They apparently do this on the basis of many items of information they are hardly aware of. I assume that trained nurses and physicians are often able to assess the condition of their patients in a similar way. •

• *Example 5.* A friend of mine was continually plagued by an eczema. A physician who was consulted did not manage to uncover a cause. My friend set out to do some research of her own. She was aware of factors that may cause eczema's, such as ingredients of food, clothing, cosmetics, drugs and the like. She made an inventory of changes in life style that might have accompanied the onset of the complaints. One thing she came up with was that she had recently started using an expensive cosmetic. So she decided to discontinue the use of the cosmetic. The eczema disappeared shortly afterwards.

This does not really prove that the cosmetic was the culprit. However, it seems to me that a sound case had been made. Notice that the example involves the elaboration of a hypothesis by inductive reasoning in the style of Mill (see *4.3*), a test of this hypothesis via the derivation of a test implication, and an explanation on the basis of the hypothesis after confirmation. •

Perhaps you think that the subject discussed here is not typical of the life sciences because it involves humans—who show placebo effects since they possess the faculty of imagination—and because the experiments described are 'field experiments' that do not allow complete control of conditions. On either count I would disagree.

Placebo effects are not typical of humans only. Conditioning is one possible source of such effects. Suppose your doctor has provided you with apparently effective medicines on various occasions. It is conceivable that this will help you recover through a conditioned response on the next occasion. That is, previous experiences may lead to the expectation that the medicines you get will work. The expectation itself may become a health-promoting factor. Now conditioning is possible in animals as well. Therefore they could show placebo responses. As illustrated in the next example, experiments have indeed confirmed this.

• *Example 6.* Ader (1985) gives the following example. There is a strain of mice with a particular auto-immune disease, that is a disease in which the immune system attacks the body it is supposed to protect. The disease can be suppressed by a drug. It has been shown that if mice with the disease get sugar pills together with the drug during a certain period, they will afterwards thrive on sugar as the sole 'drug'. Ader concludes from this that sugar induces a placebo response after a process of conditioning. •

The fact that controlled experiments on effects of drugs are 'field experiments' does not make them special either. It is true that researchers in biology and biomedicine perform many laboratory experiments which are less problematic in some respects. However, *understanding processes in the field is a very important goal of biology. Research aiming at this goal is as tricky from a methodological point of view as investigations concerning medical treatments.*

Exercises

6.4.1. Homeopathic drugs are supposed to be effective at high dilutions. At times homeopaths work with dilutions which are so extreme that no trace of the substance diluted will be left in most preparations. It is assumed that the preparations can be effective none the less, because the solvent takes over curative properties of the homeopathic drug. This is supposed to happen during the dilution process, which has special features such as shaking the preparation in certain ways.

Reilly et al. (1986) have apparently shown by a double-blind experiment that a particular homeopathic drug, at extreme dilutions, alleviates hayfever complaints. What's your reaction to this?

6.4.2. John has felt a bit ill during a long period. A physician he consults decides to run a couple of laboratory tests. The results show that John has diabetes. John is advised to take insulin shots. So he does. His condition improves soon afterwards. Would it be reasonable to conclude that the treatment with insulin is responsible for John's recovery?

6.4.3. Some medical treatments cannot be evaluated with double-blind experiments. Can you provide reasons for this?

CHAPTER 7

Laws and Theories

7.1. Preliminaries

Suppose we attend a party which suddenly goes awry as the butler of our host is shot dead. Everybody saw the butler crumple, nobody noticed the killer. The police are called in and an investigation is started. How did the butler die? That's the question to be answered.

A physicist at the party who is a bit out of touch with ordinary life comes up with an answer. The butler was hit by a bullet with a high velocity. Naturally the bullet penetrated the body. Theories of physics tell us that it should. The butler's heart happened to be in the way of the bullet. Naturally the heart stopped pumping. Theories of physics and biology tell us that it should. So the butler died. Theories of biology imply that persons will die when their hearts stop beating.

The theories mentioned by the physicist are probably acceptable. They do explain the event since they lead to the identification of a cause, indeed a chain of causes, which was sufficient to produce the butler's death. However, this will not be the cause we are interested in. In view of the context of interest the physicist's story is irrelevant.

After a thorough search the police come up with the theory that John did it. Mud on John's shoes is similar to mud found under a table near the door. A gun in John's bedroom proves to be the fatal weapon. So here is the story of the police. John had been hiding under the table. He had the gun with him. When nobody was looking in his direction he aimed at the butler, pulled the trigger, and slipped out. This theory explains the event by the identification of different causes which also played a role. It is more obviously relevant in the present context.

The second theory is the better one in view of our purposes. Both theories explain how the butler died. Our interest is in the more specific question who did it. This question is only addressed by the second theory and the explanation it involves.

Let's retain the example but change the context. We are trained investigators engaged in scientific research concerning causes of death in humans. In that capacity we are asked to account for the butler's death. Now the first explanation will do better than the second one. How did the butler die? He was shot through the heart. This explains his death, since we can appeal to the general law that all people who are shot through the heart will subsequently die. The law in its turn can be inferred from two other laws: if a bullet goes through a person's heart, then the heart will stop beating; if any person's heart stops beating, then they will die. Also, we can expand the explanation by laws describing the behavior of bullets that hit bodies. So we have a scientific theory consisting of interrelated laws which explains the butlers death.

I do not want to imply that theories of science must always consist of laws (see below). However, they should certainly be *general* to some extent. The story of the police, however true and appropriate in its context, is not a scientific theory because it emphasizes very specific things, as it should. In daily life we may describe the story as the theory that John did it. In science we would not use the term 'theory' in this way.

So much for the butler. The example is a good starting point for reflections on the concepts of law and theory, the subject of this chapter. Explanation, another important issue in the example, will be considered in the next chapter.

The example illustrates two important methodological points. *First, the adequacy of theories depends on the context. Second, the meaning of 'theory' also is context-dependent. The two points are interconnected. Different kinds of theory are adequate in different contexts. Accordingly the notion of theory can take on different meanings. Theories of science will often differ from theories we are content with in daily life. Very different kinds of theory exist also within science.*

Concerning science, philosophers used to defend the stringent view that theories must always consist of interrelated laws of nature. The idea was that theories develop from general hypotheses subjected to critical tests. Such a hypothesis, if well-confirmed, can be accepted as a law provided certain other conditions are met; some of these will be discussed in the next section. An important condition is that the hypothesis must not be an isolated statement. The criterion of *coherence* calls for systematic relationships with other well-confirmed hypotheses. Thus we arrive at the view that theories are interrelated sets of laws.

As a matter of fact, *there appear to be few laws in a very strict sense in biology and biomedicine* (see Schaffner, 1980, 1986; Van der Steen and Kamminga, 1991). In physics the situation is different. There one has the laws of Newton, Faraday, Ohm, and so forth.

Some philosophers have concluded from this that biology is not an autonomous science (see for example Smart, 1963). They hold that biology consists of applied physics and chemistry with natural history added to it. The natural history bit would not be real science. Biology apparently has no theories of its own since theories should consist of laws.

I regard this as a matter of terminology. Theories without laws may be interesting and important. One can choose not to call them scientific. That would not change much. It is true that no scientist would regard the 'theory that John did it' in the example as a theory of science. However, there is quite a spectrum of possibilities between theories as specific as this and the most general ones found in science. It would be unwise to give science a place only at one end of the spectrum.

Accordingly I will use the term 'theory' in a broad sense. Concerning the concept of law I will also be a bit more generous than most philosophers.

In the foregoing I have concentrated on one *methodological criterion*, generality. Many other criteria must be considered in the evaluation of scientific theories. Important ones are clarity, empirical content, realism, testability and confirmation by tests, consistency, coherence, simplicity, explanatory power and predictive power. In the face of this, we can but expect that theories are a mixed lot. *It is implausible that theories of the life sciences could maximally satisfy all methodological criteria at the same time.*

Methodological criteria are often at cross-purposes. Thus we may have to choose between a theory which is general but not very realistic, and one which attains realism by a sacrifice of generality. *The relative importance attached to particular criteria will depend on the context, for example the purposes we have.*

Realism will be a most desirable feature of theories used to calculate amounts of drugs which patients need. If such theories cannot be general at the same time, so be it. Theories designed to explain the spread of epidemics will be different. Many factors affect epidemics. To arrive at manageable models factors with minor effects must be disregarded. This results in general theories which are not quite realistic.

In the present chapter I will not discuss all the methodological criteria I mentioned; many of them are considered in other chapters. Here I concentrate on generality and coherence.

7.2. Generality

Generality is a composite criterion. I will reserve the term '*generality*' for the opposite of a particular kind of specificity. The statement that all birds are homeotherms is more general (less specific) than the statement that all sparrows are homeotherms because it *applies to more entities*. Generality in this sense is an important aim in science. The terms 'generality' and 'specificity' are also used in a different way. The statement that all birds are homeotherms with relatively high body temperatures is more specific than the statement that all birds are homeotherms. Yet the two statements apply to the same entities, birds. The first statement is more specific in the sense of more informative than the second one. In science we often aim at specificity, rather than 'generality', in *this* sense. To avoid confusion I will not use terms in this way.

Another common meaning of 'generality' is *'universality'*. In the philosophy of science statements are called universal if they contain a universal quantifier and no existential quantifier, and do not refer to particular places, times or individuals. Many philosophers and biologists would not regard any statement containing a species name ('chipping sparrow', *'Paramecium caudatum'*) as universal, since they regard species names as proper names for individuals. On this view a species is an individual entity occurring in a particular region in a particular period. For analogous reasons, statements about other *taxa* (genera, families, orders, etc.) will not be regarded as laws of nature.

Lastly, 'generality' is also used for *'general validity'*. A statement is not general in this sense if it is true only for a restricted domain, that is for a subset of the entities it refers to. We will accept a statement as generally valid only if confirmation over its entire domain indicates that it is realistic. Notice that validity in the sense intended here is different from validity ascribed to arguments (see chapter *3*).

The term *'law of nature'* is mostly reserved for empirical statements which one believes to be generally valid, if they in addition satisfy the criteria of generality, universality and coherence.

Laws in this sense are scarce in biology and biomedicine. In my opinion this does not warrant a negative view of these sciences. They are rife with valuable natural history, theory not composed of laws in a strict sense.

Notice that *I do not use the term 'history' in its primary sense*. As the next example shows, the study of temperature adaptation is a typical example.

• *Example 1*. Temperature is one of the most important factors in the environment of animals. It affects chemical reactions and thereby has a marked influence on almost any physiological process. One would expect there to be laws which express relations between temperature and the rates of processes in animals, but as it turns out animals show many different kinds of reaction to temperature. Here I will consider reactions to extreme temperatures. (For more details, see Cossins and Bowler, 1987; some philosophical aspects were earlier discussed in Van der Steen, 1981; the present example is reproduced with some modifications from Van der Steen and Kamminga, 1991.)

Animals can adapt physiologically to extreme temperatures in two basically different ways. Some adaptations, called resistance adaptations, change the ability of animals to withstand lethal effects of extreme conditions; the rate of vital processes may remain unaffected in this case. Other adaptations, called capacity adaptations, modify the rate of vital biological processes so as to minimize negative effects of harsh conditions. I will briefly review capacity adaptations. Here we must again distinguish two basic categories, genetic and non-genetic adaptations.

Non-genetic adaptations involve changes in the performance of an individual during its lifetime. They often take the form of compensation, but there are exceptions. The normal situation is as follows. If an organism is exposed, say, to cold, its rate functions will initially decrease. This may affect performance in a negative way. After a while, however, compensatory reactions occur which restore the original rates. Such a secondary response is called acclimation. In ideal acclimation, original rates are indeed attained again. If the compensation is less than perfect, acclimation is called partial. There are also other possibilities. Compensation may take the form of an overshoot (supraoptimal acclimation), or the response may be in the opposite direction (inverse or paradoxical acclimation).

How is all this to be explained? In many cases, responses can be understood as an adequate reaction to the ecological context. Predators which are exposed to cold will still need energy for catching prey. Indeed they will need more energy for keeping warm if they are warm-blooded. So they had better improve on the rate of metabolism which is normal for low temperatures. However, the result of this strategy may be disastrous when food is scarce. In that case a reduction of metabolism, perhaps accompanied by low body temperature and inactivity (for example hibernation) may be more adequate.

Even in one ecological context, the options realized may be very different. There is a limit to possibilities for changing one's physiology. Some features of organisms will act as constraints, and the constraints need not be the same in different species. For example, some species may not have the equipment to withstand low body temperatures. Their option may be to mobilize energy by weight reduction under cold and food scarcity. And so on and so forth.

Biologists will mostly be content with a description of temperature adaptation patterns for particular species and/or constraints and/or ecological settings. The example shows that there are no simple general, universal laws for rates of physiological processes as a function of temperature. Now it may be possible to formulate more complex laws which take account of exceptions, but there are limits to the complexity which is acceptable. I think we would have to formulate *very* complex generalizations to accomodate exceptions in the present example. As yet, the study of relations between the physiology of organisms and the thermal environment yields natural history rather than theory in a strict sense

Note. In this example I have been referring to functional explanations in an informal way. The subject of functional explanation is explicitly discussed in the next chapter. •

Notice that various methodological criteria are at odds with each other in the example. Generality and universality could have been attained only if complexity were taken for granted. The biologist's preference in this case has been to sacrifice generality and/or universality in favor of simplicity.

I have described hypotheses and laws as empirical statements, and theories (in a strict sense) as combinations of laws. This must be qualified. Some theories of science are not meant to be empirical in a strict way.

When a scientist studies complex systems she will often need to simplify. This may result in a 'theory' which describes purely hypothetical situations. 'Theories' of this kind, which are often called models by scientists, can be quite useful if real situations are approximated to some extent.

I use the term model sparingly because it has many different senses.

Example 2 shows what kind of laws we may expect in a theory which covers hypothetical situations. It also has a bridge function because it paves the way for the subsequent discussion of the so-called semantic view of theories.

Let me present a brief preview. The issues are complicated. To understand them you will have to know about *the distinction of logical and empirical statements*, which I hope is a familiar theme by now. An important point in the example is that 'laws' and 'theories' which describe hypothetical situations, are perhaps best reconstructed as logical statements. This flies in the face of the methodological criterion of empirical content. (See *6.1*: hypotheses must have empirical content; so must laws, by implication.) However, *you can choose to use the notion of theory in a different way*, such that the criterion does not apply.

• *Example 2.* Suppose we want to elaborate a law which describes the growth of populations. The simplest situation we can imagine is one in which the following assumptions are satisfied. (i) Birth rate and death rate (the number of births and the number of deaths per individual per time unit) are constant. (ii) Other factors such as migration do not influence the number of individuals in the population. These assumptions imply that population growth is described by the following differential equation: $dN(t)/dt = a.N(t)$. In this equation $N(t)$ represents the number of individuals at time t, a is a constant. The left hand side represents the instantaneous increase or decrease in numbers of individuals. The solution of the equation is $N(t) = c.e^{at}$. This is an exponential equation. Because the variable time occurs in the exponent, the equation implies that growth will continually accelerate if numbers increase (birth rate exceeding death rate). If numbers decrease, extinction will be approached with an ever decreasing rate.

Should the equation be regarded as a law? Let us notice first that growth in real populations will normally fail to satisfy the equation. If that were not so, then any single species which is on the increase would soon fill the earth. If the equation is taken to mean that all populations exhibit the pattern described it would represent a statement which is so manifestly false that it will not count as a law. However, no biologist would subscribe to this interpretation. Biologists would rather regard the equation as a formula which describes a hypothetical situation; as such it is not true or false. We can choose to use the term 'law' for equations with this function. Likewise for 'theory' as an interrelated set of such 'laws'. There is nothing against this as long as we realize what the terminology stands for.

There is yet another possibility to describe the situation. We will get a true statement which is general and universal if we take the assumptions mentioned above into account, as follows. 'As long as any population satisfies assumptions (i) and (ii), its growth is described by the equation $N(t) = c.e^{at}$.'

Let me confess that, in preparing this example, I thought for a moment that this statement is an excellent law in a strict sense of the term. However, the statement does not qualify, because it has no empirical content! The assumptions imply, logically so, that the equation truthfully describes growth. Contrary to my initial intuition (yours too?) the statement I formulated does not tell us much about the real world. It is a logical statement. Of course we could again choose to be generous with the label of lawhood, and stipulate that certain logical statements qualify as laws.

All this does not mean that the equation considered here is useless. The challenge is to specify situations in which real populations do satisfy the assumptions and the equation, and to formulate new equations for different situations. Theories dealing with relevant situations will probably not be general and universal. •

Philosophers of science have developed various abstract views on the nature of scientific theories. The most prominent ones are the received view, which few researchers accept nowadays, and the semantic view (for a discussion, see Sloep and Van der Steen, 1987). Here I will only comment on the semantic view in relation to problems with universality in biology.

The *semantic view* portrays theories as collections of *ideal systems.* (One could also use the term 'model' here.) The equation in example 2 would describe such a system. Together with equations describing other patterns of population growth it would constitute a theory of population growth.

According to the semantic view, theories cover empirical matters in an indirect way only, through *theoretical hypotheses* which say that some *empirical system* behaves in accordance with an ideal one (see Giere, 1979, for an accessible survey, and Thompson, 1989, who applies this view to biology in a lucid way). An example is the hypothesis that bacteria belonging to one species, if kept at low densities in Petri dishes under certain standard conditions, will obey the equation in example 2. The example below is worked out in more detail.

• *Example 3*. I always try to maintain a vegetation in my garden which is diverse and more or less natural. One of the plants I am fond of is *Geranium pyrenaicum*, which has many flowers throughout the summer. The species thrived in the garden in the first few years after we bought our house. Then it started to decline. I noticed that the patches it occupied were different from year to year. It always grew on patches which had been bare in early spring. This, I thought, could well explain the decline. As the vegetation developed over the years it grew more dense so that bare patches became rare. So I decided to remove all vegetation from some patches in early spring. That indeed is where *Geranium* did well in summer.

Suppose I manage to describe what happens in my garden by a mathematical equation which relates the abundance of *Geranium* in summer to the percentage of area without vegetation in spring. This would not be a law because a particular species is mentioned. Further, the equation will not be generally valid. For example, *Geranium* may be outcompeted elsewhere by other species colonizing bare patches of soil, species that do not occur in my garden. If we aim at validity, we will have to be content with the non-universal and highly specific statement which relates the abundance of *Geranium* to vegetation cover *in the garden of WS*.

On the semantic view, however, we could describe the data by a law in spite of all this. Thus we could stipulate that a universal and general statement concerning the abundance of plant species in relation to vegetation cover is a law which describes the behavior of an ideal system. Empirical matters considered here would be described by a (non-universal, specific) theoretical hypothesis which says that *Geranium* in the garden of WS behaves in accordance with the ideal system. This hypothesis would be so specific that it is not very interesting from a scientific point of view. However, further research might lead to more encompassing hypotheses. Perhaps we will be able to specify conditions which single out all empirical systems that satisfy the law. That would provide us with a universal theoretical hypothesis (a law in the primary sense of the term). Notice that theoretical hypotheses are empirical statements. Therefore the general statement at the end of example 2 would *not* qualify as a theoretical hypothesis. •

In the example, empirical issues (information concerning my garden) were described in two different ways. The first description amounts to natural history; no law is available. The second description is in line with the semantic view; it appeals to a law. At first sight this appears to be contradictory. Have we laid our hands on a law? Yes and no. The contradiction dissolves once we realize that the term 'law' takes on a new meaning under the semantic view. We are dealing with different concepts of law. The empirical issues remain as specific as they were under the new description. The value of the semantic view is not that it can provide us with laws, but rather that it deals with idealization as an important feature of science.

Beatty (1980) associated problems of universality in biology with a defense of the semantic view. Some philosophers have argued that biology compares poorly with physics since it has few (universal) laws and theories. Beatty's line of reasoning was that under the semantic view the problem of universality does not exist, because laws are now statements that characterize ideal systems.

On this view, patterns of temperature adaptation (cf. example 1) are well described by universal laws *without empirical content*. Different laws, for example laws describing various patterns of acclimation, could be formulated for different ideal systems. Empirical issues would be addressed by theoretical hypotheses saying that the behavior of some empirical system corresponds with that of a particular ideal system. As already indicated, examples 2 and 3 can be treated in a similar way. Elsewhere (Sloep and Van der Steen, 1987) I have argued that problems with universality in biology are not really solved in this way.

Under the semantic formalism the problems are rather shifted to the *application* of universal theories, that is to theoretical hypotheses. Thus the classical requirement that theories must be universal is most naturally construed as the requirement that theoretical hypotheses must be universal. Now, if it is difficult to formulate universal empirical theories in biology, it will be equally difficult to formulate universal theoretical hypotheses. My attitude would be that it is better to accept that pervasive universality is unattainable in biology.

Universal theoretical hypotheses, if possible, will have the following form: 'If an empirical system E has features F_i, then its behavior conforms to that of ideal system I'. We can also imagine theoretical hypotheses stating that members of a particular taxon behave in accordance with some ideal system. Such hypotheses would not be universal (they would contain a universal quantifier, but would mention a proper name). My hunch is that in many cases the conditions that must be satisfied if an empirical system is to conform to an ideal one, will not be fully known. If *nothing* is known about these conditions we will have to be content with existential statements saying that some empirical systems conform to a particular ideal system. My comments are only meant as an amendment of the semantic view as portrayed in the philosophy of biology.

I regard the semantic view as a useful device since many theories and models in biology conform to the semanticist's ideal systems. The full complexity of biological systems cannot be covered by them. Biologists are therefore forced to simplify.

For example, biologists may model competition and predation by differential equations for two-species interactions which disregard effects of spatial and temporal heterogeneity. In reality there will be interactions among many species, and there will be effects of heterogeneity. The equations describe what happens under conditions that are never realized. They are useful none the less because (i) they may hold approximately when deviations from these conditions are slight and (ii) they are a good starting point for investigating causes of marked deviations. Thus the use of simplifying models could help one to elaborate a list of factors that make coexistence of potential competitors possible.

Beatty's views were taken up by Stephen and Krebs (1986) in the context of optimal foraging theory. The theory has been controversial for a long time (Stephen and Krebs are among the defenders). This is caused at least partly by the absence of a fully articulated methodology (see Haccou and Van der Steen, 1992). Thus generality and universality are seldom distinguished. The researchers mostly use one label, 'generality'. Stephen and Krebs, in passages reproduced in the example below, are relatively clear about this. They appear to be discussing universality.

• *Example 4.* In the last few decades many biologists have done research on optimal foraging. This research is motivated by the following line of reasoning. Animals need to gather food by foraging. They can do this in various ways. Evolutionary theory suggests that foraging behavior will be performed in such a way that fitness is maximized. In many situations rate-maximizing foraging, specifically foraging which maximizes energy intake, will maximize fitness. So the behavior of animals will often conform to rate-maximizing models.

The biologists Stephens and Krebs (1986), who were inspired by Beatty's views on universality, comment as follows on models of optimal foraging theory.

> Optimality models fail the 'physics test' because they do not specify the range of their own validity. The models of Chapters 2 and 3 say what rate-maximizing foraging should be like, but they do not say that all animals meeting criteria X, Y, and Z will be rate maximizers, as a physical theory would. [In the terminology of the semantic view, Stephens and and Krebs are here referring to theoretical hypotheses.] Optimality models specify types of systems, and as such they are yardsticks against which to compare nature; they are not claims about what nature must be like... . ... [therefore] evolutionary biologists may have to reconcile themselves to a science that does not fit together in the tidy way they suppose physics does (pp. 212-213).

Stephens and Krebs are obviously arguing here that foraging behaviors cannot be covered by universal statements.

> Many ecologists would agree that herbivore diets, especially generalist herbivore diets, are much more complex than the diets of other consumers: rate-maximizing might (it is argued) explain carnivore diets, which are made up of prey with approximately the right balance of nutrients, but herbivores often feed on abundant low-quality prey and face the problem of selecting a balanced diet and not just maximizing the rate of energy gain ... (p. 116).

This passage deals with the issue in a more implicit way. I guess that the authors would not endorse the universal statement that all animals whose food has the right balance of nutrients will be rate-maximizers. Other passages indicate that many different factors may be responsible for a deviation from rate-maximizing. Thus it is unlikely that the universal statement is true. As the following passage implies, territorial defense is one of the factors which may interfere with rate-maximizing.

> Dynamic foraging models usually focus on trade-offs ... . For example, the following dynamic trade-off problems have been studied: conflicts between feeding and territorial behavior [and other types of conflict]. ... For example, ornithologists have observed that male great tits do much of their territorial defense in the morning, but they are also likely to be hungriest in the morning ... . A model built to study the trade-off between feeding and territorial defense might try to predict the time course of feeding and defense (p. 161).

Stephens and Krebs do not characterize the logical nature of statements concerning foraging which *are* possible. It seems to me that, in many cases, we will have to be content with existential statements saying that the behavior of some species under some conditions which are not fully known conforms to a particular model. •

The examples I gave show that there is much natural history in biology. Nonetheless universal, general theories may be possible in some cases.

At the present time it is unclear how far biology may get in attempts to elaborate universal, general theories. One reason for this is that biologists seldom fully articulate the logical form of their theories. This may explain biased pictures of biology one comes across in the philosophy of biology.

Some philosophers have suggested, for example, that theories in many areas of biology cannot be universal since they deal with particular species. The following example shows that this is an unduly pessimistic view which misconstrues the nature of biological theories.

• *Example 5.* Rosenberg (1985) holds that "... there are in biology at most two bodies of statement that meet reasonable general criteria for being scientific theories. These will be the theory of natural selection and [perhaps] ... general principles of molecular biology" (p. 219).
   He offers the following argument to show that laws are impossible in other areas of biology.

> ... all those special branches of biology, ecology, physiology, anatomy, behavioral biology, embryology, developmental biology, and the study of genetics are not to be expected to produce general laws that manifest the required universality, generality, and exceptionlessness. For each of them is devoted to the study of mechanisms restricted to one or more organism or population or species or higher taxa and therefore is devoted to the study of a particular object and not to a kind of phenomenon to be found elsewhere in the universe. [In physics the situation is differ-

ent because physics deals with] natural kinds, whose instances can and do occur throughout the galaxy. Particular species do not recur, and they cannot be expected to be much like particular lines of descent either elsewhere on the planet or elsewhere in the universe (p. 219).

Rosenberg's line of reasoning is misleading. One of his crucial theses is that mechanisms studied in various branches of biology are "restricted to one or more organism or population or species or higher taxa". He concludes from this that theories concerning mechanisms will have to mention names of particular taxonomic entities. Now it is possible that some mechanism does occur only in one species due to the fact that it has particular features which other species lack. However, this would be compatible with the universal statement that *all* organisms with these features have the mechanism. Physiologists may be able to formulate universal theories concerning the conduction of impulses in the nervous system. The fact that some organisms do not have a nervous system would not make the theories nonuniversal. It is possible that Rosenberg's conclusion is correct, but his argument is inadequate. •

Exercises

7.2.1. If you are getting cold hands outdoors as the winter approaches you may start wearing gloves. If you don't do that you may notice that, after a while, the cold doesn't bother you any more. This is an example of a phenomenon called physiological adaptation. Under repeated exposure the response in this case consists of a more rapid increase of blood flow in the hands which counteracts cooling. One wonders if this can be generalized. The phenomenon could be covered by the following general statement. Repeated exposure of an organism to adverse conditions induces a response that decreases negative effects. Or, in different words, repeated exposure of an organism to adverse conditions leads to physiological adaptation. How would you characterize the status of this statement? Could it be a law of nature?

7.2.2. Optimal foraging models have been criticized on the ground that they are not testable. Stephens and Krebs (1986) comment as follows on the criticism.

> This criticism arises because optimality modelers adopt the following procedure. A design-constraint hypothesis is erected and compared with observations; if the observations do not support the hypothesis, ... assumptions are modified in a new optimality model. "Surely," the critic says, "it is unscientific to keep shoring up the cracked facade of optimality with a scaffolding of *ad hoc* modifications; instead, one should entertain alternatives, such as the trait under study being of neutral selective value and therefore not being designed for anything (p. 207).

One of the counter-arguments presented by Stephens and Krebs is as follows.

> ... many critics confuse *ad hoc*-ism with the refinement of hypotheses. The rules of the scientific method require that a hypothesis be abandoned when it is disproved, but it trivializes the scientific method to claim that all the elements used to arrive at the hypothesis must also be abandoned (pp. 207-208).

What conclusions does this passage allow concerning the logical form of hypotheses envisaged by Stephens and Krebs?

7.3. Coherence

The criterion of coherence demands that theories consist of interrelated rather than isolated statements.

My formulation is a bit vague since interrelations can take many different forms. I would like to leave it that way since scientific theories do come in diverse kinds. I do not know of any scientific theory which does not satisfy the criterion, broadly construed. This suggests that there is no reason to discuss it in more detail. However, *various philosophers have defended* more specific, *strong versions of the criterion of coherence* which call for appraisal.

Specifically, it has been argued in the past that the laws in a theory must form *a hierarchy of deductive relationships*. In addition to this, some philosophers have maintained that theories must connect various *levels of organization* (cf. the series atom, molecule, cell, tissue, organ, individual, population, and so forth). The alleged rationale for this is *the idea that phenomena should be explained by underlying mechanisms*, that is, mechanisms at lower levels. As the following example shows, this view may well apply to some theories.

• *Example 1.* The consumption of certain plants, for example peyote, induces hallucinations. This can be explained by the presence of particular chemical substances in the plants. The explanation would refer to a statement with the form 'If substance S is consumed (in certain quantities), then hallucinations will occur'. Let's regard this statement as a law. Hallucinations are brought about by specific processes in the nervous system. Indeed the law could be deduced from two other laws, 'If substance S is consumed, then processes P will occur in the nervous system' and 'If processes P occur in the nervous system, then there are hallucinations'. Thus we arrive at a mini-theory which contains three interrelated laws. One of them is connected to the others by a deductive relationship. The theory unites various levels of organization. It connects hallucinations with an underlying mechanism. •

Some theories do have a deductive structure, others don't. Theories dealing with reactions of animals to temperature have a less rigorous structure, and they do not contain laws (see 7.2, example 1).

Deductive relationships can also exist among theories. Deduction is called reduction if entire theories are involved. In the past, philosophers of science have defended a position called reductionism, which regards pervasive reduction as an ideal for science as a whole. In an ideal, unified science physics would be the ultimate foundation. Psychology would be reduced to biology, biology to chemistry, and chemistry to physics. The ideal of a unified science, thus conceived, is not accepted any more. However, reduction and (new forms of) interdisciplinary integration are still very much on the agenda.

The classical account of *the reduction paradigm* is Nagel (1961). According to Nagel, a theory T_1 is reduced to a theory T_2 if two requirements are met. The terms of T_1 must be defined with terms of T_2 and all the laws of T_1 must be deducible from laws of T_2. Reduction in this stringent form may be achieved for theories which are in the same domain. In Nagel's terminology, homogeneous reduction may sometimes be a realistic option. Heterogeneous reduction, which would connect theories in different domains (specifically different levels of organization) is seldom, perhaps never, achieved in science if Nagel's requirements are to be met.

Accordingly various authors have weakened the requirements, for example by allowing for a correction of theories that makes deduction possible (see for example Schaffner, 1967). However, even weak forms of heterogeneous reduction are totally impossible in the life sciences, since central concepts in theories concerning higher levels of organization cannot be defined with concepts of lower level theories alone.

Discussions concerning reduction in biology have often concentrated on genetics, one of the best candidates for reduction. At first sight classical genetics has been reduced to modern, molecular genetics. Even in this case, however, we will have to face the definition problem I mentioned. The relationships between classical and modern genetics are not tight at all. They could not be accommodated by any reduction model developed in philosophy (see e.g. Hull, 1974, 1981; Falk, 1990). I will not elaborate this since the example of genetics is complex.

As example 2 indicates, limits to possibilities for reduction are more easily charted in other areas of biology.

• *Example 2.* Consider the concept of predation, which no one would like to miss in ecology. Let us assume for the sake of argument that every instance of predation could be described in lower level terms, by specifying what happens in sensory organs, muscles, nervous systems, and so forth, of the organisms concerned. The descriptions would have to be very complex. More importantly, they would be very different in different instances. Therefore it is not feasible to specify defining features of predation which derive from descriptions of lower levels of organization, that is, levels below that of individuals. The idea that we could deduce equations describing interactions between predators and preys from lower level theories is preposterous. •

Reductionism is an extreme position. Those who have opposed it have at times succumbed to alternative views which are equally extreme. For example, the thesis has been defended that biology is more fundamental than physics because all laws apply to organisms, while only a limited number of them apply to physical systems. Hence, the argument continues, physics should be reduced to biology rather than the other way around, if there is to be reduction in any form. The well-known paleontologist Simpson (1964) is among the defenders of this view. One of his arguments is considered in the example below.

• *Example 3.* I will analyse Simpson's argument in detail. My starting point is the following passage in which Hull (1974) summarizes the argument (the comments I have differ from those of Hull).

> The preceding analysis of reduction [which assumes reduction of biology to physics] has not gone unchallenged. G.G. Simpson (1964), for example, has argued that biology and not physics "stands at the center of all science" because "*all* known material processes and explanatory principles apply to organisms, while only a limited number of them apply to nonliving systems." The various branches of science are better organized "not through principles that apply to all phenomena but through phenomena to which all principles apply". Hence, if anything is going to be reduced to anything, it will be physics to biology (p. 4).

I would reconstruct the argument up to the last statement as follows.

Premise 1. If laws $L(X)$ apply to things studied by science X, laws $L(Y)$ apply to things studied by science Y, and $L(Y)$ is a subset of $L(X)$, than X is more central than Y.
Premise 2. Laws $L(B)$ apply to things studied by the science of biology, laws $L(P)$ apply to things studied by the science of physics, and $L(P)$ is a subset of $L(B)$.
Conclusion. Biology is more central than physics.

'X' and 'Y' in this reconstruction are variables, 'B' and 'P' are constants. The argument constitutes a valid deduction, but that's about the only point in its favour. I have two objections.

First, premise 1 sounds like a definition for a special concept of 'centrality'. One should realize that the conclusion therefore has a quite specific meaning which is not captured by the way it is worded. It remains to be seen if 'centrality' in the intended sense has much relevance for reduction. I will come to that.

Second, the seemingly innocuous expression 'apply to things' is tricky. To begin with, it is unclear whether one should read '*all* things' or '*some* things'. The following consideration is even more important. Would the second law of thermodynamics apply to organisms? We could give two answers to this question. (i) No, the law applies to closed systems, and organisms are open systems. (ii) Yes, the law can be cast in the following form: 'If *x* is a closed system, then ...'. In this form it is true of organisms, albeit in a trivial manner. The two interpretations involve different meanings of 'apply to'. On interpretation (i), the second premise and the conclusion of Simpson's argument are quite obviously false. But they are also false on interpretation (ii). On this interpretation Mendel's laws—indeed *all* laws—can be reconstructed in such a way that they apply to, say, molecules!

Concerning the connection between centrality and reduction, my hunch is that Simpson must have had the following argument in mind.

Premise 1. If science X is more central than science Y, then Y is reducible to X.
Premise 2. Biology is more central than physics.
Conclusion. Physics is reducible to biology.

This is again a valid argument. As shown above, the second premise is false, but let's forget about that. The following objection is more telling. Let's assume that 'reduction' has the classical meaning in this argument. How could physics reduce to biology under the classical definition? The only reconstruction accounting for this which I have been able to invent is as follows. Let's regard $L(B)$ as the laws of biology, $L(P)$ as the laws of physics. Now Simpson assumes that $L(P)$ is a subset of $L(B)$. So we can deduce $L(P)$ from $L(B)$ simply because we can deduce $L(P)$ from $L(P)$! This reconstruction suggests that Simpson's thesis concerning reduction may have been motivated by an unusually broad definition of 'biology' which would make the thesis totally trivial. I do not see how other meanings of 'reduction' could improve on this.

The example shows that we should be aware of pitfalls associated with ambiguity of seemingly innocuous words (see also chapter 2). •

In literature on interdisciplinary work known to me the emphasis is generally on benefits. Obstacles are mentioned, but researchers mostly do not consider the possibility that integration may at times be undesirable, or the possibility that some allegedly integrative theories may represent pseudo-integrations. To redress the balance I will consider these possibilities in the next section, which focuses on the role of concepts in interdisciplinary work (see also Van der Steen, 1990b and 1993a).

Exercises

7.3.1. In some countries many people suffer from sickle cell anemia, a nasty disease characterized by the presence of abnormal hemoglobin (hemoglobin S) and deformed red blood cells. The genetic factors involved in the disease are known. Persons who are homozygous for the hemoglobin S allele are likely to die of the disease. Researchers have long been puzzled by the persistence of the disease in areas where it occurs. One would expect it to be eliminated by natural selection. The solution of the puzzle is that heterozygous persons who have the normal and the abnormal hemoglobin allele appear to be better protected against another disease, malaria. The evidence shows that the phenomenon of 'heterozygote superiority' is responsible for the persistence of sickle cell anemia.

The example involves phenomena at two 'levels of organization', that of individuals and their features (their phenotype), and that of genes. Parenthetically, the levels involved here do not belong to the range I distinguished above (from atom to population and beyond). Genes can be studied at the level of populations!

The information summarized here represents a coherent body of knowledge which I would gladly give the title of theory. Would it be possible to reconstruct the theory of sickle cell anemia as a universal theory? Does the theory have much generality? Is the theory the result of reduction in view of the fact that it involves various 'levels'?

7.3.2. It has been argued that theories about higher levels of organization must be reducible to theories about lower levels, because entities at higher levels are composed of entities at lower levels. Criticize this argument.

7.3.3. The criterion of coherence is introduced by Hull (1974) in the following passage.

> Two issues seem implicit in most objections raised to the commonest examples of biological theories and laws. First, many of the examples of laws ... are not process laws; that is, laws that permit the inference of all past and future states of the system, given the values for the relevant variables at any one time. ... Even if one grants that laws other than process laws can count as genuine scientific laws, a second major difficulty remains—distinguishing between natural laws and descriptive generalizations. It may be true that none of the first seven presidents of the United States of America were Christians ..., but this is hardly a law of nature. ... Numerous criteria have been suggested to mark this felt difference. However, the only one that shows any promise of being adequate is the actual or eventual integration of natural laws into theories, while descriptive generalizations remain isolated statements (pp. 70-71).

In a different passage (p. 75), Hull applies the criterion to ontogenesis. He argues that individual steps in biochemical sequences that characterize development may well be covered by laws. But what about the sequences as a whole?

> What is needed before we can count statements of such developmental sequences as laws is the recognition of some system of laws from which they can be derived. To do this, we must enlarge our area of interest to include the genotype of the organism and how it functions to control this particular pathway (p. 75).

According to Hull it is difficult to elaborate theories of ontogenesis in this way.

Do you agree with Hull's application of the criterion of coherence to ontogenesis?

7.4. Concepts Revisited

A researcher wants to elaborate a unified theory of everything. His starting point is that all processes we can think of involve change. In view of this the researcher decides to define a concept of evolution as follows. 'System X shows evolution' $=_{df}$ 'system X changes in the course of time'. The technical concept of moving factor is introduced for factors which are responsible for evolutionary change. 'Y is a moving factor of system X' $=_{df}$ 'Y is a cause of evolutionary change in X'. The stage is now set for the formulation of the following law of nature. 'All evolutionary processes are caused by moving factors.' The researcher happily concludes that he has uncovered a law of nature which explains all processes one can come across in the universe.

Of course you will not buy any of this. The reasons are obvious. First, the researcher's definitions of concepts are very broad. An advantage of this is that they apply to many different things. The price he pays is that they say very little about anything. Second, the 'law' is a logical statement. It has no empirical content since it is true by virtue of the definitions given. Therefore it cannot explain anything.

Researchers do use broad or ambiguous concepts in attempts to elaborate integrative theories. In many cases it is anything but easy to evaluate the result, because advantages and disadvantages of integration can be weighed in different ways.

Many integrative theories are indeed controversial. I will give two examples of integration involving ambiguous or over-general concepts. Next I will present a longish example of a putatively general theory that has no empirical content due to inappropriate definitions.

• *Example 1*. Evolutionary biologists and sociobiologists have elaborated various theories to account for altruism. 'Altruism' in this context has to do with reproductive (dis)advantages. You are behaving altruistically towards someone if your behavior increases her probability of reproductive success and decreases yours. The biological concept of altruism is complex; I won't repeat details (cf. *2.3*, example 2).

It has been suggested by a few authors that the elaboration of sound biological theories of altruism would make theories developed by social scientists and ethicists superfluous. Such biological theories would be very general. Some representatives of this extreme view have a tendency to amalgamate the biological notion of altruism with the concept that is current in social science and ethics. That is an unfortunate move since 'altruism' in social science and in ethics has a meaning which has nothing to do with reproduction. Moreover, 'altruism' in these fields is intrinsically connected with intentions, a subject largely outside the scope of biology.

A more modest view is that biological theory might conceivably cover certain aspects of altruism in the ordinary sense which have to do with reproduction. However, we should realize that these aspects will not be the ones social scientists and ethicists are primarily interested in.

The bold view I mentioned envisages a theory which is much more general. The generality is realized through inadmissible ambiguity (for further details, see Voorzanger, 1984, 1987). •

• *Example 2*. Bonner (1980) has articulated a general theory of culture. According to him culture is by no means restricted to man.

Bonner defines 'culture' broadly, as follows.

By culture I mean the transfer of information by behavioral means, most particularly by the process of teaching and learning (p. 10).

'Information', in the definition, also has a broad meaning.

Biology may have something to say about culture in *this* sense, so Bonner, as a biologist, is not trespassing on foreign ground when he attempts to develop an evolutionary theory of culture. But biological theories of culture need not yield important insights into aspects of 'culture' which sociologists and anthropologists like to consider.

Even if we stick to biology, the danger is that *concepts* like 'culture' are stretched so as to cover many different phenomena. This could wrongly suggest that they are a good basis for developing integrative general *theories* of culture. I am skeptical since vagueness and ambiguity tend to increase as concepts become more general. Some forms of generality are an asset, many are not. •

• *Example 3*. The concept of stress plays an important role in attempts to develop general and, in some cases, interdisciplinary theories (see *2.5*, example *4*). In my view the concept gets an unduly heavy load in various areas of science, specifically plant ecology and psychosomatic medicine. Here I will discuss plant ecology; the example is reproduced with modifications from Van der Steen (1990a). Extensive comments on 'stress' in psychosomatic medicine are given in Van der Steen and Thung (1988).

In plant ecology, stress plays an important role in a theory about evolutionary strategies developed by Grime (1979). According to Grime, plant strategies are mainly determined by levels of stress and disturbance in the environment. He distinguishes three viable strategies. Species representing them are characterized as competitors (low stress, low disturbance), stress tolerators (high stress, low disturbance) and ruderals (low stress, high disturbance). I will focus on stress and stress tolerance (for other key concepts, and a more detailed analysis, see Van der Steen and Scholten, 1985).

Stress is defined by Grime as "the external constraints which limit the rate of dry matter production of all or part of the vegetation". In his theory, the quantity R_{max} (potential maximum rate of dry matter production in a standardized productive environment, measured in the first phase of the life cycle) plays a leading role as a measure of production. This means that stress, as a feature of the *environment*, is characterized through features of *plants* such as a low R_{max}. How could we characterize stress tolerance as a feature of plants? Again a low R_{max} seems to be an appropriate indicator, and Grime appears to use it in this way.

This conceptualization of stress and stress tolerance is not wrong in itself, but we should be aware of one important consequence. Consider the general statement that stress tolerators occur in environments which represent stress, which Grime would probably regard as an important empirical truth. Indeed the statement is true, in virtue of the fact that stress and stress tolerance are characterized by similar indicators. It is not empirical since it tells us nothing about nature. Grime aims at a general empirical theory which relates features of plants such as stress tolerance to features of the environment such as stress. However, his treatment of concepts precludes the elaboration of such a theory.

The comments given so far are informal and they need further justification. To make my criticism stick, I will present a more detailed analysis which you can skip if you are not interested in technicalities. For the record, Grime thinks my analysis is totally wrong; plant ecologists I know who oppose Grime's views have reacted favorably.

Grime primarily applies his (very informal) conceptual scheme to whole vegetations, but he uses it for species as well. Only species will be considered here. Grime's definition of 'stress' allows various interpretations. Consider the following versions, in which E is a symbol for environments, and S for species.

D1. 'E_1 represents stress relative to E_2' =df 'production of S_i in E_1 is lower than production of S_j in E_2' (S_i are species characteristic of E_1, S_j are characteristic of E_2).

D2. 'E_1 represents stress for S relative to E_2' =df 'production of S is lower in E_1 than in E_2' (S stands for any particular species occurring in both environments).

D3. 'E represents stress for S_1 relative to S_2' =df 'production of S_1 is lower than that of S_2 in E' (E stands for any particular environment containing the two species).

Now suppose we try to establish relations between environments and strategies. What result would we get under the various definitions?

D1. Suppose some E_1 characteristically has species S_i with a low production (in E_1), and E_2 has S_j with a high production (in E_2). On Grime's approach, R_{max} is a criterion of production in either case. If S_i and S_j are suitable indicators, E_1 will represent more stress than E_2. S_i occurs in E_1, so it should have properties characteristic of stress tolerators. This 'prediction' is indeed borne out since S_i has the lower production in terms of Grime's R_{max} criterion. However, the prediction is a logical statement. The predicted property cannot fail to materialize since the *definition* ensures that it will be present if S_i occurs in a stressful environment. Of course this will not do. Any hypothesis about relations between environments and species so degenerates into vacuity.

D2. This definition, however 'natural', leads to a different problem. If it is adopted, it is impossible to establish relations involving stress between species and environments if Grime's R_{max} is used as a criterion of stress tolerance (supposedly a feature of species). R_{max} would call for a comparison of species in one and the same environment, whereas the definition of 'stress' can only be implemented by a comparison of environments. The two things simply don't mix.

D3. Suppose some S_1 has a lower production than S_2 in some E. Then E will represent more stress for S_1 than for S_2 (lower production). And S_1 will presumably be a better stress tolerator *because of this* since R_{max} as a criterion of stress tolerance concerns similar differences in production. Again we are threatened by vacuity because stress and stress tolerance are defined in similar terms.

In Grime's theory, the concept of stress apparently has a heavy load, it covers features of species and their environment in one stroke. As a result, the status of relations between species and their environment is obscure. Grime's research is intended to uncover *empirical* relations. At the same time, a *logical* link is forged by the meanings given to central concepts. Thus the empirical content of the theory becomes unclear. The remedy should be obvious. It would be wise to abandon the logical link. Species and environments will have to be characterized in independent terms. Production is a good feature to characterize species, so environments had better be described in different terms.

It will not be easy to find a common denominator for environments which *prima facie* represent stress. So general statements about effects of stress may need to be replaced by more specific statements which have little in common, statements about poor soils, heat, drought etc. For example, the statement that plants on poor soils, specifically soils with a low nitrogen content, have a low R_{max}, may well represent an important truth of plant ecology. I doubt whether attempts to elaborate a more general theory of 'stress' make sense at all. •

Exercises

7.4.1. Computers have memories. Does that imply that computer science can account for the phenomenon of memory in man?

7.4.2. In recent years some ethologists have argued that biology should be concerned with the study of mental phenomena in animals. Griffin (1984) has proposed that we must try to develop a cognitive ethology. An evolutionary approach could provide more unity in research on the mental. How could we explain consciousness in animals in evolutionary terms? Griffin gives the following answer. The adaptive value of consciousness should be obvious. Conscious thought is a simple, economical tool which can greatly improve planning and prediction. Very complex mechanisms would be needed to achieve the same results without consciousness, and evolution will foster the development of simple tools.

Criticize Griffin's argument.

7.4.3. Over the centuries, the mind-body problem has perhaps been the hardest problem faced by scientists and philosophers alike. Mental phenomena and physical phenomena seem to belong to entirely different categories which cannot be covered by a single theory. Bunge (1980) is among the philosophers who have recently tried to elaborate an encompassing theory. He defends a position he calls *psychoneural identity theory*, or *emergentist materialism.*

After a general outline of his approach, Bunge first introduces key notions and principles concerning brain functions. He then tackles mental states and processes. Mental processes as a general category are explicitly defined by him in terms of specific physiological processes in the brain (p. 74). Bunge subsequently shows that the definition of the mental in these terms has far-reaching implications, for example: "All mental disorders are neural disorders" (p. 75). How could such an approach be justified? Bunge presents the following arguments.

> We have not characterized mental states and events independently of brain states and events, as is usually done. We have not proceeded this way for a number of reasons. Firstly, because mentalistic predicates, such as 'sees red' and 'thinks hard' are, though indispensable, coarse and vulgar rather than exact and scientific. Secondly, because the whole point of the neurophysiological approach to mind should be to get away from the ordinary knowledge approach and render mind accessible to science. Thirdly, because, if mental events are characterized independently of brain events, then the identity theory turns out to be either idle or false ... . Fourthly, because some physical events too might be given (simpleminded) descriptions in everyday mentalistic language, and so would be 'proved' to be mental (p. 80).

Do you agree with Bunge's view?

7.5. Afterthoughts

Science does aim at the ideal of generality and universality. Examples in this chapter show that this ideal can only be realized to a limited extent in the life sciences. Indeed one should not expect anything else.

The idea that scientific theories could satisfy all methodological criteria which merit our attention is preposterous. There will have to be priorities. The priorities we set will depend on the purposes we have with a theory; priorities are context-dependent.

Generality and universality will have to compete like all other criteria. There is no warrant for the assumption that they will always win. Hence we should expect there to be theories in the form of natural history.

The life sciences do contain much (valuable) natural history. The study of natural history has not been high on the agenda in the philosophy of science. We urgently need a methodology of natural history.

We might be tempted to allot more force to some other methodological criterion, one which theories must always satisfy. Empirical content is an obvious candidate. Science is supposed to tell us something about the real world, hence its theories should be empirical. This seems to be an eminently reasonable view, but we should be cautious even here.

The term 'theory' is used in various senses. Under the semantic interpretation, theories do *not* have empirical content. Semantic view theorists locate the empirical content of science elsewhere. The force of the criterion of empirical content for theories obviously depends on the way we reconstruct science. What kind of reconstruction is the most adequate? That is a question without a general answer. The value of any particular reconstruction will depend on the purposes we have with it. Context-dependence again!

I have argued that some attempts to achieve interdisciplinary integration in the life sciences are flawed and ill-conceived. *Philosophers have often concentrated on examples of successful integration. However, there are failed integrations as well, and integration may well be undesirable in some cases.*

Conceptual analysis is indispensable for the evaluation of putative integrations. At times concepts in science are given a heavy theoretical load. This may result in pseudo-integration.

Philosophers are aware of the fact that concepts of science are loaded with theory. Less attention has been paid to possibilities for manipulating the load of concepts (see also 2.6). We should aim to develop a methodology that indicates what kinds of manipulation are fruitful in the elaboration of scientific theories.

Some of the examples I gave are meant as an outline for such a methodology.

CHAPTER 8

Explanation

8.1. Explanations as Arguments

If you touch a hot stove you feel pain. Why would you be in pain? If anybody asks this question they will presumably be searching for an answer that constitutes an explanation. They may get divergent answers depending on the person whom they ask.

The simplest answer is that exposure to heat with a certain intensity always causes pain. This explanation associates a particular occurrence with a generality concerning occurrences of a certain kind. A natural way to understand how the explanation works is by giving it the form of an argument. We can deductively infer your feeling pain from the premise that exposure to heat always causes a feeling of pain, which we may regard as a law, and the premise that you were exposed to heat.

A neurophysiologist would give a different explanation. She would describe processes in receptors in the skin which respond to heat and processes in the nervous system which cause pain. Electrical impulses from receptors will reach the brain through the spinal cord and stimulate certain areas responsible for the experience of pain. This explanation could also be given the form of an argument which appeals to a generality, various generalities in fact.

Which of the two explanations is the better one? The second explanation is obviously more informative, it is an expansion of the first one. However, that need not make it more appropriate. Perhaps we will be satisfied with the first explanation because our interest is in the cause which it identifies. The value of any particular explanation will depend on the context of interest.

Yet another explanation concentrates on functions rather than causes. Why did you feel pain? Because it is useful to have pain in certain situations. The pain in the example is indeed useful. The reason is not that it makes you withdraw your hand. You will do that before you feel the pain. The pain does have a function

125

since it reminds you that it is wise to nurse the afflicted hand and be careful with it. That will promote healing. This explanation is different from the other ones in that it does not refer to causes, it is a so-called functional explanation. Some researchers would not accept it for this reason. Functional explanations are controversial.

The following example shows that the patterns of explanation I distinguished are by no means the only possible ones. Suppose we set out to explain the occurrence of a particular car accident. Many explanations will be feasible. One, the driver hit the tree because he was drunk. Two, the tree happened to be in the car's way when the driver swerved to the right. Three, the road was slippery. Four, the brakes of the car were not in an optimal condition. And so forth. All these explanations mention a cause that contributed to the accident. Each of them may be acceptable in some context. The police will be interested especially in the condition of the driver and the condition of the brakes. The city council may be interested in trees which should be removed because they decrease safety. All the explanations refer to a cause retrospectively identified. They do not seem to involve generalities with the status of law. It is unlikely that any of them can be reconstructed as an argument with premises implying that the accident occurred. Yet the explanations appear to be acceptable.

Do explanations identify causes? Do they contain laws? Do they allow of reconstructions that transform them into arguments? My answer to such questions would be, sometimes they do, sometimes they don't. The category of explanations, like that of theories, is heterogeneous, in daily life but also in science.

Philosophers have elaborated various models of scientific explanation (for a historical survey, see Salmon, 1989). At times their work seems motivated by the desire to arrive at *the* philosophical model of explanation. As yet there is no generally accepted model (cf. diverse views in Kitcher and Salmon, 1989). I think there never will be any, precisely because we are dealing with a heterogeneous category.* However, this by no means implies that the philosophical models are useless. If they are not generally valid we can choose to apply them to contexts which they do cover. That is the strategy I will follow.

---

* I have had some confusing discussions about this with US philosophers of biology. They kept saying that my arguments are unconvincing. In retrospect I think we were using the term 'model' in different senses. Of course *one* encompassing model may describe the *heterogeneity*.

To begin with I introduce a classic model which I think is still quite useful, although it has well-known shortcomings, the model of Hempel and Oppenheim.

In a famous article, Hempel and Oppenheim (1948) introduced the idea that explanations are *arguments*. The conclusion of an explanatory argument, the *explanandum*, describes the phenomenon to be explained. The premises, the *explanans*, consist of statements permitting the derivation of the explanandum. In accordance with this idea, I did construe various examples of putative explanations as arguments (see especially *3.2*, examples *1*, *2* and *4*).

An explanation which fits this description is not thereby adequate. According to Hempel and Oppenheim, explanations must satisfy the following conditions of adequacy (methodological criteria).

1. The explanation must be a valid deductive argument.
2. The explanans must contain essentially at least one general law.
3. The explanans must have empirical content.
4. The sentences constituting the explanans must be true.

The example below makes Hempel and Oppenheim's criteria more concrete.

• *Example 1*. 'Birds which consume food containing pesticides (of a certain kind, in certain amounts) produce eggs with thin shells.' This statement could qualify as an outline of a general law. If observations show that the shells of the eggs in a particular robin nest are thin, we could explain this by appealing to the law. The explanation has the following pattern.

$$\frac{(x)(\text{if } Bx \text{ then } Ex)}{Ea}$$

'*B*' stands for 'is a bird (or pair of birds) which has consumed food containing pesticides', '*E*' means 'produces eggs with thin shells'.

The argument represents a valid deduction, the premises have empirical content, and they may well be true. All in all the explanation fits Hempel and Oppenheim's scheme rather nicely. •

The example illustrates that laws alone do not permit the deduction of statements about particular facts. In addition we need statements concerning these facts in the premises. These are called *initial conditions*.

Laws themselves can be explained as well. If a law is the target of an explanatory argument, the premises will only contain (different) laws. Examples 2 and 3 illustrate this.

• *Example 2*. The thesis that organisms containing chlorophyll (a pigment that plays a role in photosynthesis) do not occur on planets without oxygen in the atmosphere may qualify as a law. The law can be deduced from the more general law that there are no organisms containing chlorophyll in places without oxygen. •

• *Example 3*. It is always possible in principle to expand an explanation by explaining premises it contains with a new argument. Concerning the explanation in example 1 one wonders *why* ingestion of pesticides (of a certain kind, in certain amounts) leads to thin egg shells. Suppose the pesticides inhibit a particular enzymatic reaction, and that this leads to thin egg shells. This would explain the first premise of the original explanation in the following way ('I' is used for 'is a bird in which enzyme function ... is impaired').

$$(x)(\text{if } Bx \text{ then } Ix)$$
$$\underline{(x)(\text{if } Ix \text{ then } Ex)}$$
$$(x)(\text{if } Bx \text{ then } Ex)$$

In this argument a law is explained on the basis of two other laws. •

Some philosophers would not regard the explanation in example 1 as satisfactory because it does not refer to underlying mechanisms. The explanation merely subsumes a special case under a regularity. I would regard this as a matter of terminology. One may choose to use the term 'explanation' in broad or in narrow senses. Arguments by subsumption are anyway part and parcel of science.

The expanded explanation in example 3 does refer to an underlying mechanism. Needless to say, the process of expansion could be carried on much farther.

Hempel and Oppenheim's model turned out to have shortcomings. For one thing, some explanations are not readily cast in the form of an argument with the explanandum in the conclusion. However, I would not simply reject the model for this reason. Why not apply it to explanations which *can* be reconstructed as arguments? After all, we need methodological criteria to evaluate the arguments involved. Now the criteria, as formulated by Hempel and Oppenheim, have also been criticized.

The first criterion, that explanations must be valid deductive arguments, is too stringent because some explanations can only be construed as probabilistic arguments. Hence the model must be extended (as Hempel later did) to accommodate such arguments.

The fourth criterion, that the explanans must be true, also needs to be modified. We can never be sure that universal premises in an explanans are true since universal statements cannot be verified. Therefore the criterion should be replaced by the requirement that explanans statements must be true or at least well-confirmed.

It seems to me that the third criterion, which demands that the explanans has empirical content, is reasonable. The second criterion, that explanations must contain a law, is more problematic. Consider the following example.

• *Example 4*. Suppose you ask a biologist why chiffchaffs in temperate areas are migrants, and he replies that all insectivorous birds in these areas are migrants. Would that constitute an adequate explanation? Let's have a look first at the logical structure of the explanation.

$$(x)(\text{if } Ix \text{ then } Mx)$$
$$\underline{\qquad Ia \qquad}$$
$$Ma$$

Here 'x' is a variable for species, 'a' is a constant representing the chiffchaff, 'I' and 'M' stand for 'is an insectivorous bird species in a temperate area' and 'is a migrant', respectively.

Many philosophers would balk at this example since they would not regard the first premise as a law. Well, should it count as a law if it is true? It is difficult to answer this question since philosophers do not agree about the explication of the concept of law. Some hold that laws must express causal relationships. If we accept that criterion the first premise is not a law, since being an insectivore does not cause migration.

I would regard the argument as a potential explanation, although its explanatory force is admittedly weak. To some extent, I think, the issue is a terminological one. We can *choose* to use the terms 'law' and 'explanation' in broad or in narrow senses. Nothing much hinges on this as long as it is clear what terminology is being used. •

I will evaluate explanations which can be reconstructed as arguments with *methodological criteria* already discussed in chapters *3* and *4*: *clarity, validity* (in the case of deductive arguments), *confirmation of premises, and coherence.* I use the criterion of coherence as a heterogeneous closing entry which, in different contexts, can be specified in different ways.

If an argument is meant as an explanation of a particular phenomenon, it will have to meet the additional criterion that the explanans must contain a premise which is more general than the explanandum.

My criteria resemble those of Hempel and Oppenheim, but they are weaker and to some extent more informal. The next example illustrates the way I would use the criterion of *coherence*.

• *Example 5*. Consider an explanatory argument with the following form.

$$(x)(\text{if } Bx \text{ or } Px, \text{ then } Ex)$$
$$\underline{\qquad Ba \qquad}$$
$$Ea$$

As in example 1, '*B*' and '*E*' stand for 'is a bird which has consumed food containing pesticides', and 'produces eggs with thin shells', respectively; '*P*' means 'is an old car'.

This is decidedly an odd explanation. However, it satisfies almost all the criteria I have listed. To reject the explanation, I would appeal to the criterion of coherence. The inclusion of '*Px*' in the argument does not affect its validity, and it does not change the truth of premises. '*Px*' is irrelevant in the sense of redundant. Also, the connection between '*P*' and '*E*' does not cohere with available knowledge. •

Much philosophical work on explanation has gone into attempts to defuse odd 'explanations' such as the one in the last example. It is extremely difficult to exclude them all, even by models of explanation which are more formal than the one I have presented. As an exercise in philosophy, attempts to exclude them are important. The ensuing models, however, are often cumbersome, so that it is difficult to apply them in science.

Odd counter-examples of explanations which apparently invalidate philosophical models are convincing as counter-examples because it is obvious that they represent bad science. That is why scientists need not attend to all the niceties of philosophical models.

The simple array of criteria I have introduced may not fit all the bills which philosophers think should be paid. Whenever they don't we will have to supplement them with common sense. The next example illustrates some discrepancies between explanations as covered by philosophical models and 'original' versions.

• *Example 6.* Suppose we know John has a fever, due to a bacterial infection, that will probably continue for some time in the absence of treatment. John is prescribed an antibiotic, but contrary to expectation the fever does not abate on the next day. We could explain this by the assumption that he harbours a strain of bacteria which is resistant to the antibiotic. The following argument describes the situation.

$$(x)(\text{if } Bx \text{ and } Ax \text{ and } C, \text{ then } Fx)$$
$$\underline{Ba \text{ and } Aa \text{ and } C}$$
$$Fa$$

'*B*' stands for 'is infected by bacteria (which cause protracted fever in the absence of treatment)', '*A*' for 'receives an antibiotic'; '*C*' expresses the condition that the bacteria are resistant to the antibiotic, '*F*' means 'has a fever after one day', '*a*' represents John.

This reconstruction does not wholly capture the point of the argument since it does not express the fact that *C* covers an exception (cf. 'contrary to expectation'). The following reconstruction is more to the point.

$$(x)(\text{if } Bx \text{ and } Ax, \text{ then } Fx \text{ if and only if } C)$$
$$\underline{Ba \text{ and } Aa \text{ and } C}$$
$$Fa$$

The first premise now in addition implies that the fever will go down if the bacteria are not resistant to the antibiotic.

The last reconstruction still leaves some aspects of the explanation unexpressed. Thus the material implication does not express a causal relationship. In fact no formalization will be able to cover all aspects of verbal arguments. Things left out in a formal reconstruction must be taken care of by inspecting the context in an informal way.

There are no fixed rules for finding the most adequate reconstruction. The choices you make will be a function of the things you want to be clarified. •

The examples presented above show that we often need to reconstruct explanations to uncover their logic. The reconstruction of texts from live science will mostly involve more than that.

In science texts phases of research such as hypothesis testing and explanation often do not receive explicit labels, and they may be interwoven in a single train of thought. Reconstructions should take this into account.

Here is an example.

• *Example 7.* One of the most important theses of homeopathy is the so-called similia principle. According to this principle, a substance qualifies as a drug for a disease if it produces symptoms of the disease in healthy persons. The principle is due to Samuel Hahnemann (1755-1843). In a publication which defends homeopathy, Coulter (1984) describes the first phase of Hahnemann's work as follows.

> ... in the 1790s he discovered a new interpretation of "similarity." Knowing that quinine was curative in malaria, he decided to ascertain its effects on a healthy person. He took a strong dose himself and soon started to exhibit the typical symptoms of malaria. He concluded that quinine acts curatively in this disease because of its capacity to elicit malarial symptoms in a healthy person ... (p.58).

Coulter subsequently states that many tests afterwards confirmed the similia principle.

Most scientists think that the tests performed so far are inadequate. I will not discuss this point but concentrate on what's happening in the passage quoted.

According to Coulter's portrayal of Hahnemann's reasoning, the similia principle emerged from an inductive argument concerning quinine and malaria. The curative effect of quinine is at once explained by him in terms of the principle. No illegitimate circle is involved here provided the principle is tested in independent ways.

The explanation has the following form.

$$(x) \text{ (if } Pxy \text{ then } Qxy)$$
$$\frac{Pab}{Qab}$$

In this scheme 'x' and 'a' are a variable and a constant, respectively, for (candidate) drugs, 'y' and 'b' are similarly used for diseases, 'Pxy' means 'x produces the same symptoms as y', 'Qxy' means 'x cures y'. The argument is a valid deduction. It would be perfectly acceptable if the first premise would be well-confirmed.

Notice how much reconstruction is needed before the pattern of the argument becomes clear. •

Exercises

8.1.1. Some species of plant have idiosyncratic needs concerning nutrients. For example, there is a *Viola* species which can grow only on soils with high zinc concentrations. A biologist knows that this species is common in some area. However, he notes that it is absent in one particular locality in the area. Measurements show that this locality has a different soil with little zinc in it. Would this explain the absence of the species? If so, would the explanation involve deduction?

8.1.2. In the past, some philosophers have defended *vitalism*, a doctrine which opposes the idea that physics and chemistry suffice to explain biological phenomena. Vitalists hypothesized that phenomena left unexplained by physics and chemistry must be due to an unknown vital force. Would you agree that it may be reasonable to explain phenomena which have resisted explanation on the basis of physics and chemistry by attributing them to the action of a vital force?

8.1.3. According to the so-called principle of Gause, two species which have the same resources, and whose numbers are not affected by other limiting factors, cannot coexist in the same area. Suppose research shows that two particular species (i) coexist in some area, (ii) have the same resources and (iii) also have other limiting factors. Would the principle of Gause explain their coexistence?

8.2. Causation, Prediction and Alternative Explanations

Philosophical investigations during the last few decades have shown that many crucial methodological notions are troublesome. There is no agreement about the proper explication of 'theory', 'law', and 'explanation'. This indicates that such notions represent heterogeneous categories. Apart from this, explications which heavily emphasize a formal approach may by this very feature fail to capture all essential shades of meaning of ordinary scientific discourse.

My approach has been to keep formal approaches at a minimum, and to rely on the context for a fuller understanding.

The concept of causation, likewise, has resisted a unitary philosophical explication.

One reason for this is that the term 'cause' is used for rather different items. Also, elementary logic (see comments on the material implication in chapter *4*) cannot fully capture the nature of causal connections, and more advanced forms of logic are controversial. Yet some tools of elementary logic are useful for distinguishing various meanings of 'cause'.

The notions of *necessary condition, sufficient condition* and *necessary-and-sufficient condition* are important for understanding causation. This was already explained in *4.4*; let me recapitulate the comments given there. In logic the meanings of these notions are as follows. If some statement 'if p then q' is true, the state of affairs described by 'p' is sufficient condition for the state of affairs described by 'q' to occur. If p is realized, q cannot fail to occur. The converse is not true, but q is obviously a necessary condition of p. If q does not occur, then p will not occur either. The statement 'p iff q' represents a necessary-and-sufficient condition which goes both ways, because it implies both 'if p then q' and 'if q then p'. Analogously we can regard feature P as a sufficient condition for feature Q if the statement '(x)(if Px then Qx)' is true, and so forth.

These notions from logic provide only a partial explication of conditions which are *causally* necessary, or sufficient, or necessary-and-sufficient. We should remain aware of connotations which get lost when causal idiom is expressed in this way. In ordinary language, we will say that a condition causes an effect, if the effect can be changed by manipulating the condition. Also, effects are assumed not to precede causes. These aspects of causal relationships are not covered by the material implication.

The term 'cause' is sometimes used for necessary conditions, but it may also stand for sufficient or necessary-and-sufficient conditions. In addition there are causes in a weaker sense. Miscellaneous examples in this section and the next one will illustrate the various meanings of 'cause'.

• *Example 1*. Predation pressure of a certain intensity may cause a population to go extinct if it also has a density at a particular low level. Predation and low density are causal conditions in this case. A fire might also cause extinction. The fire may represent a sufficient condition. Likewise for the combination of predation and low density. However, a particular predation pressure alone could be neither necessary nor sufficient. Likewise for density at a particular low level. •

If a causal factor is sufficient for a particular effect, we can deduce the presence of the effect from the presence of the cause. That is, from the statement '(x)(if Px then Qx)' (here taken to represent a causal connection) and the statement 'Pa' (here taken to represent the causal feature P of a) we can infer the statement 'Qa'. Conversely, if a causal factor is necessary for an effect, we can deduce the presence of the cause from the presence of the effect. An inference which goes from cause to effect is then impossible.

In scientific work we often aim at the discovery of particular causal factors which interest us for some reason. If the relevant factors are necessary but not sufficient (or neither necessary nor sufficient), they are none the less often labeled as 'the' cause. It is important to realize this in the interpretation of scientific work.

The identification of causal factors is often deemed explanatory, even if it is impossible to reason from causes to effects. This implies that some explanations cannot be reconstructed as arguments dealing with effects in the conclusion.

The example below is a case in point.

• *Example 2*. Suppose a doctor is faced with a patient who has a disease which is always fatal if no treatment is given. Only one treatment with a drug is known. Unfortunately the drug leads to recovery in only ten percent of the cases. Other factors which play a role in recovery are unknown. The doctor prescribes the drug, and the patient recovers.

Most people would say that the treatment was the cause of recovery, and that it explains the recovery.

Notice that the expression '*the* cause' is misleading. The treatment is apparently one of the causal factors influencing recovery; other factors are unknown. In common parlance, both in science and in daily life, the expression 'the cause' is frequently used for reference to one of the causal factors we happen to be interested in. In the present example, the cause is a necessary condition which is not sufficient.

In what sense is the recovery explained by the treatment? One thing is obvious. It is reasonable to infer in this case that the treatment was among the causes of recovery. We can construct a satisfactory deductive argument to this effect. In many cases scientists will indeed say that they have explained a phenomenon if they have managed to identify a causal factor in this way. I regard this as a legitimate usage of the term 'explanation'.

The explanation does not conform to the Hempel-Oppenheim model. It involves an argument which moves from effect to cause, but inferences in the opposite direction fail.

The most natural way to represent an inference of the latter kind would be as follows (see chapter *4.3*):

$$p(R|T) = q$$
$$\frac{Ta}{Ra} \quad [q]$$

This is a probabilistic argument in which '*R*' stands for recovery and '*T*' for the treatment. The value of *q* is 0.1. If we replace '*R*' by 'not-*R*' in the argument, the value of *q* will become 0.9. On the basis of available information, therefore, it is more reasonable to infer that the patient does not recover.

Now we can supplement the argument with the information that the patient did in fact recover. Other conditions necessary for recovery must have prevailed. This line of reasoning is represented by the following argument.

$$(x)(\text{if } Tx \text{ and } C, \text{ then } Rx)$$
$$\frac{Ta \text{ and } C}{Ra}$$

In this argument '*C*' stands for unknown conditions, whose presence is inferred from the event described by the conclusion. If the inference is not supported by any additional information, as in the present case, the argument does not have much force since it is an epistemic circle. It is in fact a scheme for an explanation which might develop into a substantive explanation in the future, if we discover more about *C*. •

The explanation in the last example amounts to the identification of a cause. It is obvious that not any cause will do. The big bang is a remote cause of the recovery in the example, yet it does not explain it. *Which causes are relevant?* This question can often be answered in a natural way by construing the explanation as an answer to a *contrastive why-question* (see Van Fraassen, 1980; Lipton, 1991; chapters *3.4* and *4.3*).

A particular patient, let's call him John, recovered from a nasty disease. Why did John recover? Because he received a particular treatment. If we give this answer, we are probably contrasting the phenomenon to be explained with an alternative. A possible alternative we may have in mind is a hypothetical situation in which John would have died for lack of treatment. Thus we allot a causal role to the treatment because it is the factor which makes a difference. The question addressed here turns out to be contrastive: Why did John recover instead of dying? It may also stand for a different contrast: Why did John recover whereas Mary died? If Mary did not get the treatment, the answer should point again to the causal role of the treatment. However, if Mary died in spite of receiving the treatment, we will have to search for a different cause.

The role of contrasts is particularly prominent in explanations which attribute a disease—or some other feature of an organism—to a genetic or an environmental cause.

All features of organisms are caused by genetic *and* environmental factors. Hence the mere thesis that a disease is genetic is a poor basis for an explanation. The thesis is meaningful only if it is taken to mean that the difference between persons with the disease and healthy persons can be accounted for by a genetic difference (see 2.5, example 2).

In 7.2 I argued that biology has few laws in a strict sense of the term.

The mere complexity of living organisms often precludes the formulation of laws. To reduce complexity biologists often elaborate 'laws' which describe the behavior of hypothetical or ideal systems. This feature of biology is emphasized by the semantic view of theories, which construes theories as collections of ideal systems.

At first sight, theories in this sense are a poor basis for explanation because they say nothing about the real world. However, as the following example shows they can be very useful as a standard of comparison in contrastive explanation.

• *Example 3.* The simplest models of population genetics deal with alternative forms of a gene ('alleles') which are associated with differences in fitness, in the sense of expected reproductive success. If one allele is clearly superior, the other allele will disappear from the population if certain conditions are satisfied. In real populations things are seldom that simple. Suppose we come across a particular population in which two variants of a gene persist in spite of differences in fitness they produce. That would call for an explanation. The question to be answered is, why do the variants coexist? One factor that might account for this is migration (there are many more possibilities). Perhaps the variant with the highest fitness has a more pronounced tendency to migrate. This would justify the explanation that migration causes persistence. The explanation is contrastive; it involves a comparison of the real population with a hypothetical one. We will allot a causal role to the factor or factors which make the difference. •

In the wake of Hempel and Oppenheim's paper many philosophers have defended *the thesis that every explanation corresponds with a prediction.* The idea is that there is no difference between explanatory and predictive arguments. From the premises of an argument we can infer an event after it has occurred (explanation) or beforehand (prediction); the argument will be the same in either case.

Example 2 shows that *this line of reasoning does not apply to all explanations and predictions.* If we can explain an event only by an argument which moves from effect to cause, prediction will be impossible.

There are other exceptions. In some cases premises of an explanatory argument which moves from cause to effect can only be identified once the effect is known to have occurred. (This need not amount to circular reasoning; see *3.3.*) Prediction would not be possible then.

Various philosophers have pointed out that the latter kind of example does not invalidate the thesis that explanations are potential predictions. According to them the thesis is only meant to assert that prediction would have been possible if premises would have been known beforehand. This is correct, but the important fact remains that it is not possible in many cases to know premises beforehand.

Conversely, predictions need not correspond with explanations. The behavior of barometers often suffices to predict storms, it will certainly not explain them.

In many other respects the logic of prediction is similar to that of explanation, which is why I have chosen not to discuss prediction in more detail.

In scientific research, we will often have to make a choice from a set of alternative explanations. If such a set would represent an exclusive and exhaustive classification of alternatives, we could end up with one correct explanation. In reality, however, it is seldom meaningful to classify alternative explanations in this way.

Different explanations may concentrate on different causes in the sense of necessary conditions of a phenomenon. If that is the case the alternatives will not be exclusive. The context will determine which alternative is appropriate. Different explanations mentioning sufficient conditions may likewise coexist. Examples 4-6 illustrate the issue of alternative explanations.

• *Example 4.* If a jay (I am thinking of the bird species which is common in Europe) sees a sparrow hawk it will generally utter an alarm call. We can regard the presence of the sparrow hawk as a cause which explains the alarm call. Seeing the hawk will normally suffice for the jay's call.

Would the explanation be satisfactory? Well, that depends. If we hear an alarm call and wonder what factor caused it, the information that the sparrow hawk rather than something else induced it may constitute an acceptable explanation for us. The question 'Why did the jay call?' is then taken to mean 'What factor led to the call?'. The answer is that the sparrow hawk rather than something else induced it.

The question can also be taken to mean 'Why did the jay call *at all*?' It would be safer for it to keep silent. On this interpretation the search is for an evolutionary answer. Biologists would argue that there are causal selection processes which promote the use of alarm calls in jay populations, although they are seemingly disadvantageous for individuals when they are uttering them. •

• *Example 5.* The absence of a species on a particular island could be explained in various ways. It is possible that the species originated on the mainland, and that it does not have the capabilities for dispersal overseas. Another possibility is that there is no suitable habitat for the species on the island. The explanations are compatible. Both refer to sufficient conditions. •

• *Example 6.* In the temperate zone birds in a particular category, insectivores mostly, do not breed in winter. We could ascribe this phenomenon to low temperatures and short photoperiods. Alternatively, we could argue that low concentrations of certain hormones will preclude breeding in winter. Both explanations would correctly point to causes which are sufficient conditions in this category of birds. The causes mentioned form a 'chain' in this case. Physical conditions cause changes in hormone production, and these changes influence reproduction. •

The search for generality in science fosters causal language at abstract levels. At times that is rather misleading. Consider the following example.

• *Example 7.* Many different causal factors are implicated in evolutionary processes. Thus competition, predation, physical factors such as temperature and humidity, and so on, are involved in processes of natural selection. We can summarize this by saying that natural selection is the cause of evolution. That is misleading in various ways. In the first place evolution need not always involve selection. Thus 'random drift' (random processes that cannot be attributed to an environmental influence) may also cause evolutionary change. In the second place natural selection is definitely not a factor like temperature. The statement that natural selection causes evolution actually asserts that *there are* all sorts of factors, such as temperature, which cause evolution in a particular way, namely by inducing differential perpetuation of variants. Natural selection is not a cause over and above these factors.

Notice that the thesis that natural selection causes evolution, upon unpacking, appears to have an existential component besides a universal one. •

Exercises

8.2.1. Behavior can be studied in many different ways. One could argue that a full-fledged analysis of behavior must deal with underlying physiological mechanisms. Alternatively, to understand behavior in an ecological setting, one can study evolutionary history. Which alternative would you prefer?

8.2.2. Causes of diseases are called etiological factors. Such factors come in many different kinds. The only feature they have in common is that they do help to produce disease. The thesis that etiological factors cause disease is indeed a true statement. Would you regard the thesis as an appropriate expression of a causal relationship?

8.3. Special Cases

Some authors have maintained that, in the life sciences, there are special explanations which do not conform to the standard patterns discussed above, namely *functional explanations* and *historical explanations*. These categories of explanation are the subject of this section.

To understand functional explanations we will need to know the meaning of *the concept of function*. During the last few decades philosophers have elaborated various explications of 'function' without reaching consensus. Three major categories of explication have been distinguished (see e.g. Reznek, 1987, chapter 6).

According to the *evaluative account*, the function of a feature of an organism consists in activities sustained by the feature that promote 'the good' of the organism. In line with this account we could argue, for example, that the function of camouflage in some organisms is reduction of predation, since this is useful to the organism.

The *teleological account* connects functions with 'goals' of the organism rather than 'goods'. I will disregard it since it closely resembles the evaluative account.

Lastly, on the *etiological account* an activity associated with a feature is a function of the feature, if the performance of that activity has been a cause of the feature's origin, and is a cause of its maintenance. This account is the one which most philosophers favor nowadays. According to this account, reduction of predation should be regarded as a function of camouflage only if it has played a causal role in the origin of camouflage (in the evolutionary past) and is causally implicated in the maintenance of camouflage.

Philosophers discussing such accounts typically search for counter-examples to check whether they are generally valid.

Here are two examples against the evaluative account. "My nose may contribute to my good by supporting my spectacles, but this is not one of its functions. The production of heart sounds may aid diagnosis, and thereby serve the good of the individual, but this is not a function of the heart" (Reznek, 1987, p. 102).

We can always try to accommodate counter-examples by making the theory they undermine more specific. That is not something I will attempt to do here. I would rather argue that *the term 'function' represents a heterogeneous category, so that no single explication is likely to cover all the uses it has in the life sciences.*

The etiological account puts the concept of function in an evolutionary context. The function of camouflage is reduction of predation. Reduction of predation is an effect of camouflage. This effect is responsible for the maintenance of camouflage in populations through natural selection, and it will have done so in the past. Evolutionary biologists may indeed use the concept of function in this way. Physiologists, I think, are more likely to use an evaluative concept of function. If they attribute functions to features of organisms, they will normally think of how they work out ('function') in the present. If confronted with awkward counter-examples such as the ones mentioned by Reznek, they would dismiss them by an appeal to common sense. Anyhow, I fail to see why the concept of function should have evolutionary connotations in *all* contexts in the life sciences.

What about *functional explanations*? We could succinctly characterize such explanations as answers to why-questions, as follows. Question: Why does organism x have feature y (for example camouflage)? Answer: Because y performs function z (for example reduction of predation) in organism x. At first sight, effects are thus explaining causes rather than the other way around. No doubt a biologist who gives a functional explanation has something else in mind. If the *etiological* concept of function is his concern, his explanation will presumably amount to the following theses. First, he will assert that some feature does function in a particular way. Second, he will assume that we can give an evolutionary explanation of the way the feature functions. Thus the 'functional' explanation would be a normal causal explanation of a feature that enhances survival. This is illustrated by the example below.

• *Example 1.* Suppose a new organ is discovered in some species. Research on the organ shows that it produces a hormone implicated in reproduction. The reaction of many biologists would be to say that this explains the presence of the organ. This would be a succinct reaction. Let us expand it. The presence of the organ in an individual is causally responsible for the production of the hormone. Naturally, hormone production in an individual is *not* responsible for its having the organ. However, if the hormone aids in survival, it is implicated as a causal factor in the maintenance of the organ *in the population*. It may also be reasonable to assume that the hormone has played a role in the past in the maintenance, perhaps also the origin, of the organ. •

'Functional explanations' which represent an etiological account apparently do not constitute a special category of explanations. They rather are ordinary explanations of functions as a special category.

Let's go back to the *evaluative account*? Could it be a basis for explanations of a special type? I do think a case can be made for this, but let me emphasize that the issue is controversial. A referee who commented on the manuscript for this book recommended deletion of the text on functional explanations on the ground that biologists regard them as obsolete. A few days after receiving this advice I happened to chair a meeting of biologists designed to chart the varieties of explanation in biology. I expected to witness a heated debate on functional explanation. To my surprise all the participants agreed that functional explanations are special and that they constitute an important category of explanation. The following example is in line with this assumption. It captures the gist of the discussion at the meeting.

• *Example 2*. An important theme in evolutionary biology is that animals tend to show behavior that maximizes fitness in the sense of reproductive success. Biologists often explain behaviors by showing that they maximize fitness. On the etiological account of function the core of the explanation would be as follows. Behavior B of species S has effect E. E maximizes fitness. Natural selection in the past has promoted B since alternatives were less viable. Continuing selection is now responsible for the maintenance of B in S. I will not reconstruct this scheme in more detail because I want to concentrate on an explanation in line with the evaluative notion of function, which involves high fitness, or a feature associated with fitness, as a 'good'. Optimal foraging theory is an important source of such explanations (see also 7.2, example 4). My example fits this theory.

Optimal foraging theorists assume that if certain conditions—which admittedly are not fully known—are met, animals will forage in such a way that the net rate of energy intake is maximized. The underlying idea is that fitness will also be maximized by this. The rate-maximization idea could be expressed by the following 'law'. If conditions C are satisfied, and an animal has various options concerning foraging behavior, it will realize the option which maximizes energy intake. In principle, we can explain a particular behavior by a deductive argument with this law as a premise, if it is reasonable to assume that conditions C are satisfied (a big if). We will also need a premise stating that a particular option maximizes energy intake. From the two premises we can infer that the behavior conforms to this option. This looks like a reasonable explanation if the animal we observe does behave in this way. The law and the explanation are 'special' in the sense that they do not refer to causes in a direct way. (I have simplified. Actual explanations in the literature involve mathematical modeling.)

Suppose we observe foraging in great tits which need to feed young. If we put their behavior in the context of optimal foraging theory, the distribution of prey in the area surrounding the nest will be important. How can the tits maximize their own net energy intake and the energy supplied to the young? If caterpillars they feed on are abundant near the nest they should forage close to the nest. If the caterpillars are much more abundant in a patch at a greater distance from the nest, it may be wise to forage there. Flying to the rich patch costs energy, but this may be off-set by a decrease in the energy needed for searching. In principle we can determine which behavior would maximize net energy intake. If the tits do show this behavior we have arrived at an explanation which is rightly called functional.

If observations indicate that the great tits do not in fact maximize energy intake we need not therefore reject the optimal foraging approach. Instead we may assume that conditions C, for example conditions concerning predation, are not satisfied in this case. The tits may avoid rich patches if visiting them would expose them to a greater risk of predation. The crucial point is here that maximization of energy intake need not always coincide with maximization of fitness.

Notice that this explanation of deviant tit behavior is contrastive. The behavior observed is compared with a behavior that would maximize energy intake. The factor of predation explains the difference. •

It has been thought that *historical explanations* are special because they concern unique events. They have been pitted against explanations on the basis of laws. Many authors have argued that historical events cannot be covered by laws because they are unique (see e.g. Goudge, 1961). This thesis is not as clear as it seems because *the concept of uniqueness is tricky.*

Uniqueness as such is not a meaningful notion because *all things are unique in some respects.* The description we give determines whether or not an event is unique. Now defenders of historical explanation as a special category could argue that events *as described* in the study of history, for example evolutionary history, are unavoidably unique. Dinosaurs went extinct during one particular period in the course of history. This event is doubly unique. The extinction of a *particular* taxon cannot happen twice. Also, the event as described is necessarily unique since it involves a *particular* time.

It is easy to rebut the argument that such necessarily unique events cannot be explained on the basis of laws.

The following argument scheme suffices for this purpose.

$$(x)(t)(\text{if } Pxt_i \text{ then } Qxt_{i+1})$$
$$\frac{Pat_1}{Qat_2}$$

Here 'x' is a variable for things, 't' is a variable for time, 'P' and 'Q' represent particular properties, 'a' represents a particular individual. Notice that 't_i' and 't_{i+1}', unlike 't_1' and 't_2', do *not* represent constants.

The scheme shows that we can use a law, together with an initial condition, to explain an event which is described as necessarily unique because a particular point in time is mentioned.

Likewise for the dinosaurs. Their extinction was a unique event. The mere fact that it was unique does not logically preclude its being explained by a law concerning mass extinctions.

To evaluate claims concerning historical explanations we must move beyond abstract argument and see what theories about historical processes are like. In point of fact, in biology such theories and explanations based on them seldom refer to laws. I would not conclude from this that historical explanations are special qua historical explanations. Areas of biology which are not concerned with history do not have many laws either. This is due primarily to the complexity of living systems. Biologists will often have to be content with explanations resorting to moderately general statements which are not laws. Also, some explanations will merely amount to the retrospective identification of causes, in cases that do not permit arguments from causes to effects. *All in all there is no reason to allot a special status to historical explanations.*

The following example illustrates the role of explanation in history as studied in biology. It is reproduced with modifications from Van der Steen and Kamminga (1991).

• *Example 3.* The history of life is characterized by continuous processes of extinction and speciation. The rates of the processes have shown marked fluctuations. Over and above continuous background extinction there have been episodes of mass extinction on a global scale which affected a great variety of higher taxa, followed by periods of increased speciation. Mass extinctions have a great impact on life on our planet, so it would be interesting to know their causes. In principle, the extinction of a species can be caused by a great variety of factors, for example 'normal' climatic ones, biological ones such as competition and predation, and extraneous factors such as volcanism and meteors hitting the earth.

So mass extinctions need not have a common cause. There is indeed much controversy about their explanation. Stanley (1987) has recently defended the thesis that all of them have been caused by the same factor, cooling on a global scale. Stanley's thesis that global cooling is all-important is by no means shared by all geologists (see articles in Donovan, 1989; Raup, 1991), but I am not here concerned with its merits. My aim is merely to analyse his line of reasoning because it shows again that the orthodox conception of explanation on the basis of laws need not apply everywhere in science.

One of the factors which might be implicated in mass extinctions is changes in sea level. Stanley reasons as follows about this factor (see e.g. pp. 36-40). When sea levels are high, continental shelves (areas of shallow sea adjacent to continents) are extended. As levels get lower, there is a vast reduction in shelf area. Now the number of species that can live in a particular region is known to depend on the area of the region. So numbers of species, at least of animals that inhabit shallow seas, should decrease as the sea level decreases.

However, according to Stanley, changes in sea level cannot explain all mass extinction events. For example, many of these events also involved pelagic life far beyond the margin of continents. There is also a stronger argument against the sea level explanation, which involves laws, or at least empirical generalizations. The relations between diversity (species numbers) and area can be expressed by species-area curves. Now we have enough data to construct curves for terrestrial organisms, but it is more difficult to get at curves for marine organisms. We are not totally powerless, however. The data we have show that life in some continental shelves with a restricted area is diverse. So diverse, in fact, that it is reasonable to suppose that reductions in species numbers (among animals inhabiting shallow seas) which are known to occur during mass extinctions cannot be explained by a reduction of continental shelf areas alone.

The latter argument may be reconstructed as follows. Data in our possession are compatible with many parameter values of species-area curves for animals in shallow seas. But no member of the set of admissible curves could be used for explaining mass extinctions. This shows that it is possible to reject a scientific explanation on good grounds even though the precise form of the 'law' it contains is not known.

Stanley follows a similar line of reasoning with respect to many other factors. He ultimately concludes that global cooling is the best remaining candidate to explain known mass extinctions. In addition, there is positive evidence pointing to a crucial role of this factor. The most important part of Stanley's argument consists in showing that the extinctions were in fact accompanied by global cooling. This amounts to uncovering initial conditions.

What about the role of laws in the explanation? Stanley does not mention any law. But it is obvious that laws are presupposed here. Organisms have limited ranges of tolerance. So it is safe to assume that substantial, relatively rapid changes in some environmental factor should result in extinctions (if this is not a law, it surely is a sketch for a law). Would the temperature changes which accompanied mass extinctions have been marked enough to account for the phenomenon? That is a moot point. The inference is reasonable if one can show that the magnitude of the changes was well beyond commonly observed ranges of tolerance, and that their rapidity would not permit genetic adaptation on a large scale. Stanley's data show that this may well have been the case. The positive arguments for temperature as the main causative agent would be stronger if laws and initial conditions could be given a more precise form, but that is hardly feasible. But the arguments *are* strengthened by negative ones showing that other factors can be excluded on reasonable grounds. By and large, the resulting pattern of explanation may be satisfactory.

So we end up with an interesting empirical generalization (itself not a law), that all the mass extinctions which occurred on the earth in the past have been caused by global cooling. The evidence contains all sorts of elements. There is much useful inductive reasoning. Laws do play a role in the background, though they are not available in a very precise form.

In many respects, I side with those who have rejected the thesis that historical explanations are special. The example shows that in biological research the emphasis often is not on laws. Historical explanations do constitute an example of this. In many cases a weaker kind of general statement will suffice for perfectly sensible explanations. •

The example deals with an explanation which is merely historical in the sense that the events to be explained happened in the past. We should realize that past events can also be crucial in the explanation of the present. For example, the question why an organism has a particular feature may call for an historical answer. Likewise for the species composition of communities. In the last decade sophisticated methods have been developed to chart the role of 'nonhistorical' and 'historical' factors that have shaped life as we now know it. I will not discuss them since they are geared to research by specialists. If you are interested you should consult the excellent, integrative survey given by Brooks and McLennan (1991).

Exercises

8.3.1. The diversity of bird species in the US exceeds that in Europe. This has been explained as follows. Complexes of mountains in Europe often stretch from east to west, whereas the direction in the US is mostly from north to south. During the ice ages, bird species were only able to survive if they could shift their area of distribution toward warmer regions. In Europe many species failed because they could not cross mountains between cold and warm regions. In the US the birds did not need to cross mountains.

Could this be an acceptable explanation? Does the explanation involve laws?

8.3.2. The etiological conception of function may not sit well with the thesis that functions are a result of natural selection. Why would that be so?

8.4. Afterthoughts

The material presented in this chapter confirms a thesis already defended in previous ones, that *the search for highly general philosophical models of science is misdirected. Concerning explanation it is obvious that there is no single pattern which scientific work always conforms to.*

Many explanations can be reconstructed as arguments with a conclusion describing the explanandum event, but some can't. Scientists will often say that they have explained a phenomenon if they have uncovered one of its causes, even if they cannot come up with an argument that permits the derivation of effects from statements about causes. Philosophers might protest that this does not constitute 'real', solid explanation. It seems to me that this is a matter of terminology. Uncovering causes is often important. Those who are unwilling to call this explanation may choose a different name, but they should grant that we are dealing here with an important aspect of science.

If an explanation can be cast in the form of an argument, doing so will facilitate methodological evaluation. For this purpose, the methodological criteria formulated by Hempel and Oppenheim are useful, but they need to be weakened. I have indicated how that can be done. The criteria may fail in some contexts. My reaction to this is that *we should specify contexts of application rather than continue to replace criteria by ever more sophisticated ones* (see also Sloep and Van der Steen, 1988). The chances are that no set of criteria will apply to all contexts.

There are few laws of nature in the life sciences. Therefore Hempel and Oppenheim's requirement that explanations must contain at least one law is definitely too strong. Explanations will often involve general statements which are not laws.

The concept of cause and the concept of function, like that of explanation, represent heterogeneous categories that cannot be covered by a single philosophical explication.

Some functional explanations are best construed as 'ordinary' explanations of the causal variety. In addition to this there are functional explanations which definitely do not center on causes. Historical explanations are special at most in the sense that they seldom refer to laws. However, they share this feature with many other explanations in biology.

In this chapter I have again used elementary logic as a methodological tool. Elementary logic has its limitations. For example, it cannot fully express the concepts of cause and effect. However, this does not make it useless. Aspects of meaning which are lost in the process of translation can be covered by supplementary comments that take the context into account. After all, perfect translations are impossible.

CHAPTER 9

Facts and Values

9.1. Preliminaries

The applications of methodology to research in the life sciences I have presented so far concern science viewed as an isolated entity. The present chapter puts the life sciences in a broader context. Methodology should be as useful in this context as in science narrowly conceived. That is what I intend to show by examples. The emphasis will be on relations between science and ethics.

Let me introduce some informal terminology first. In this chapter I will use the term '*values*' for 'norms and values'. The expression '*normative statement*' will mean 'statement expressing norms or values'.

By and large, the methodology introduced so far will suffice. However, as I will show in section *3* it needs to be extended somewhat because we will have to deal with values and with relations between facts and values.

I have strong views about the current situation in ethics. In brief, I would endorse the following theses. First, methodology is underdeveloped in ethics; in this, ethics compares poorly with science and with philosophy of science. This leads to spurious problems and needlessly protracted controversies. Second, theorists in ethics overstress the importance of highly general theories. They should become more modest and concentrate on low-level generalities with a context-dependent validity. Third, if ethics would be reconstructed along these lines, it would come to resemble science in many ways.

These opinions have been defended in a recent paper. It is reproduced here as an appendix which substantiates my approach of values in this chapter.

To avoid misunderstanding let me say from the outset that I do not regard methodology as *the* tool with which we can solve ethical problems. Methodology helps us clarify arguments and evaluate them. That is a modest role, one which is singularly important though.

147

A broader view of sciences and normative issues is presented in chapter *10*, which is beyond the textbook format.

Throughout the book I have been at pains to *keep logical and empirical issues apart*. Some philosophers have argued that this cannot be done. I would grant that the distinction is often blurred in the practice of science. Therefore *the thesis that there is no watershed between the logical and the empirical* makes sense as a description of science. However, the absence of a watershed must not simply be taken for granted. We should aim to reconstruct science in such a way that logical and empirical statements are distinguished to the extent that this is possible.

Common sense indicates that *the thesis is fishy* anyway. This is revealed by questions which plainly indicate that there is a distinction to be made. 'What do you mean?' calls for an elaboration of logical matters. 'Why do you think so?' does often represent a demand to produce empirical evidence.

Concerning values some philosophers are also challenging classical distinctions. According to a classical view science is concerned with an objective approach of empirical matters. It is *value-free*. This presupposes that we can distinguish between *facts* (empirical matters) *and values*. Dissenting philosophers have argued that this, too, cannot be done. I would disagree with their view for similar reasons. No doubt facts and values are often amalgamated in science. That does not imply that it should be impossible to keep them apart.

The demand that science must be value-free is actually too unspecific. Values can play different roles in science. Some of them are quite legitimate. First, methodological values are an indispensable part of science. Second, science can rightly be concerned with facts concerning values. Third, the choice of a particular subject of research, and the way one performs research, will be determined by values, not least ethical values. There is nothing wrong with that. To the contrary, one can make a case for the thesis that science must be regulated by values in this way.

The idea that science must be value-free should rather be interpreted as the demand that scientific theories must be empirical. Even this needs to be qualified.

Theories cover aspects of phenomena we are interested in. Many aspects will be left out. So theories are necessarily biased in a broad, non-pejorative sense of the term. Likewise for data of science we choose to concentrate on. Any articulated perspective on phenomena is linked up with valuations in view of things left unsaid, even if all things said are empirical and true. In the rest of this chapter I will stick to the pejorative usage of the term 'bias'.

The thesis that science should be value-free makes sense only if the term 'value-free' is given a very restricted meaning.

All these strictures, however important, should not make us forget that it is important to distinguish facts and values and to aim at purely empirical scientific theories.

9.2. Distinguishing Facts and Values

In ordinary discourse we do not normally keep facts and values apart. The way we express ourselves indicates that we are able to mix them without trouble. In itself this is not a cause for concern. The things we mix need not be mixed up in the process. Mixing facts and values is not a good thing in all situations, though.

In complicated situations, or conflicts, the ways of ordinary discourse may become a stumbling block because they are not geared to analysis. Analyses will not get us very far unless we keep facts and values distinct.

The example below illustrates the way facts and values are commonly mixed in ordinary discourse.

• *Example 1.* Consider the following fictitious dialogue between two persons, *A* and *B*.

> *A* "I have discovered that John is not ill after all. He is simulating. He's afraid. I think he doesn't want to be in his office today. The boss is coming back, so ... ." [Things left unsaid.]
> *B* "I got a different impression when the doctor was here yesterday. Have you forgotten about his [John intended] temperature and the way he coughed last night?"
> *A* "I don't know. I guess he may have faked that."
> *B* "Maybe you can do that with a cough. But fever?"

A full-fledged analysis of such a dialogue (of *any* dialogue) would be quite a job. I will restrict myself to selective comments concerning facts and values.

(i) "Discovered": this wording suggests that the speaker intends to convey facts (in an autobiographical way). It indicates that we are dealing with an empirical statement (one that may well be false). However, the next sentence puts the issue in a different perspective. True, up to a point "John is simulating" also points to alleged facts, but it is linked up with a value judgement. Or is it? What would the intended flavor of "simulating" be?
(ii) "He doesn't want to be in his office today". This is an empirical statement (which may well be false) expressed by the speaker in an autobiographical way. Notice that it says something *about* values (at issue are values allegedly entertained by John). Notice further that the statement, though

empirical, may be misleading. It represents bias in view of things left unsaid (consider *B*'s response). However, this kind of bias does not turn empirical statements into normative ones.

(iii) "Have you forgotten ... ?" That's a question, not an assertion. However, the intention of *B*'s question is to remind *A* of some facts. In a reconstruction of the situation we should uncover such facts and express them with empirical statements.

(...)

(iv) "But fever?" Only a question, a very succinct one at that. But notice what's happening underneath. *B* implicitly maintains that people can't fake a fever. That's a highly general empirical statement. *B* infers from the generality that John did not fake the fever he allegedly had. This is a valid deduction. If the premise is true *B* has a strong case. •

I do not mean to suggest with the example that conversations in daily life should be analysed methodologically. If you would try that out during a party, you would soon run out of conversation for lack of partners. Methodological analysis is out of place in most contexts. In some contexts, though, it is useful or even desirable. Let me change the context a bit to make this clear.

• *Example 2.* In the past John stayed at home on many occasions, pleading illness. His boss gets suspicious, he thinks John has developed a habit of faking illness. He comes with a threat. If John is indeed faking he will get fired. The boss phones the doctor but that doesn't get him anywhere. The doctor flatly refuses to give him any information on the ground that he is bound by professional secrecy. John's partner likewise is unwilling to provide information. She doesn't believe John but she wants to be loyal. When she has the boss on the phone she is evasive. John will know best.

In a setting of this kind the persons involved will presumably feel a need to keep facts and values distinct. Is it *true* that John is faking illness? That appears to be an empirical matter. (I am assuming here that the concepts of illness and faking are clear in the context.) If it is indeed true that John is faking, what *should* I do about it? That's where values enter the scene. In the situation described, different values are important for different persons because their positions are different. That need not imply, for the rest, that their general views of values are different.

"I must be loyal to my partner. If I would tell the boss, I would not be loyal. So I don't need to tell the boss." That is an example of an argument which connects facts and values. •

Concerning the last part of the example you should notice that *distinctness* of facts and values does not preclude *connectedness*. I will not analyse any argument here; the next section deals with the analysis of arguments.

Exercise

9.2. Consider the following quotation from a news report (New Scientist, 4 July 1992).

> On the number of embryo's transferred [in test-tube baby treatments], the authority [the Human Fertilisation and Embryology Authority] believes that "there is no longer any clinical justification for transferring more than three in any circumstances". Concern centres on the risk of multiple pregnancies, which are hazardous for both mothers and babies and put enormous strains on those who raise children (p. 6).

a. Is the statement quoted within the quotation ("there is ... circumstances") an empirical statement, or is it a normative statement?
b. Is the statement expressed by the last sentence ("Concern ... children") an empirical statement, or is it a normative statement?

9.3. Analysing Arguments: The Use of Methodology

It is obvious that ethics can be brought to bear on science. Should science, conversely, be allowed to influence ethics? Here we should notice first that one kind of influence would be illegitimate.

To the extent that science is empirical we cannot derive normative statements from it. An argument which only has empirical premises, and a normative conclusion, is fallacious.

Philosophers call an argument of this kind a *naturalistic fallacy*. (Some dissenters, for example Richards, 1986, think they can beat the naturalistic fallacy.)

The term 'naturalistic fallacy' was coined by the philosopher Moore. He used it for arguments which move from 'is' to 'ought'. I am using the term for a slightly different fallacy. This usage is common in philosophy, but it is not historically accurate.

The fact that empirical premises cannot yield a normative conclusion does not imply that science is irrelevant to ethics. Arguments which have normative in addition to empirical premises do allow the derivation of (other) normative statements.

Consider the normative thesis that the use of certain pesticides must be banned. A thesis to this effect will normally be based on the assumption that the use of pesticides has certain consequences (an empirical matter) which are undesirable (a normative matter). Charting consequences is clearly a job for science.

Arguments in which empirical and normative matters are linked in this way can be evaluated with methodological criteria discussed in the previous chapters. However, the criteria need some amendation.

Normative statements are non-cognitive, or at least not wholly cognitive (see chapter *1*). That is, they can be acceptable or unacceptable, but we cannot simply regard them as true or false. This calls for an *expanded concept of validity*. An argument is valid in the ordinary sense when the conclusion cannot be false if the premises are true (see *3.1*). With respect to normative statements in an argument we will have to read 'acceptable' for 'true' and 'unacceptable' for 'false'.

The two examples below show that we can analyse arguments connecting facts and values with ordinary methodology if we use the expanded concept of validity.

• *Example 1.* Consider the thesis that organochloric hydrocarbons (a kind of pesticides) must be banned because they are persistent and therefore pose a threat to the environment and human health. In this formulation the thesis is a statement, but it is obviously meant as an argument which can be reconstructed as follows.

$$(x)(\text{if } Px \text{ then } Cx)$$
$$(x)(\text{if } Cx \text{ then } Bx)$$
$$\underline{\hspace{2cm} Pa \hspace{2cm}}$$
$$Ba$$

'*P*' means 'is a persistent pesticide', '*C*' means 'has such and such consequences upon being used ...', '*B*' stands for 'must be banned', '*a*' is a constant representing organochloric hydrocarbons.

Some logicians would not agree with my use of the expression '*Bx*'. It suggests that 'must be banned' is an ordinary kind of property, which it obviously isn't. I think this use of symbols is innocuous as long as we recognize that the second premise is a normative statement.

My formalization allows for the evaluation of validity. The canons of ordinary logic show that the argument is valid, if they are expanded with the notion of validity introduced above. For further comments see example 2. •

• *Example 2.* 'The use of penicillin must be banned, because some strains of bacteria will acquire resistance to antibiotics such as penicillin and become more pathogenic as a result.' This argument is easily reconstructed such that we get a formalization identical to the one in example 1. Thus we are dealing with a valid argument in either case. Now validity does not suffice to ensure acceptability. The arguments will have to satisfy other methodological criteria as well.

In point of fact the arguments as reconstructed here are not acceptable at all. They appear to represent a fallacy called *argumentum ad consequentiam* (see *3.3*). The fallacy consists in judging a way of action on the basis of a *particular* consequence. To reach a sound judgement we will have to consider the impact of all positive and negative consequences we can get information about.

Specifically the arguments do not satisfy the criterion that premises must be acceptable. In example 1, the premises of the reconstructed argument do *not* refer to additional consequences. The problem is located in the second premise, which is not acceptable. However, the argument could be transformed into an acceptable one by the inclusion of an additional premise which says that there are alternatives for the use of persistent pesticides, with better overall consequences.

The pesticide argument can be expanded with an acceptable clause to this effect. The antibiotics argument is different. It cannot be transformed into an acceptable argument. •

Exercises

9.3.1. 'Vivisection on invertebrates should be allowed, because invertebrates cannot feel pain.' A statement of this kind presumably will be meant as an argument. On the most simple reconstruction, the conclusion ('vivisection ... allowed') is based on one premise ('invertebrates ... pain'). Notice that it will not be easy to test the premise. Is the argument a naturalistic fallacy on this reconstruction? Is the conclusion an empirical statement? An additional premise is needed to get a more plausible reconstruction. What premise would you add?

9.3.2. Consider the following statements made by a biologist and an ethicist.

> *Biologist*: "Evolutionary biology shows that altruism is impossible because natural selection promotes egoism, that is, behavior that increases one's own fitness."
> *Ethicist*: "Altruism obviously occurs in humans. Many acts of people benefit others rather than themselves."

The two statements appear to be incompatible. Which side would you take? Specify your reasons.

9.4. Deplorable Gaps*

If you read a representative selection from the literature in ethics you won't come across much science. Real science I mean, not fabricated science or science fiction.

Believe it or not, ethics and more generally philosophy contain much science fiction. What would your world look like if you would be a brain in a vat? That sort of thing.

---

* In view of the general nature of this section there will be no exercises.

I have heard ethicists saying that doing science is the business of scientists, not ethicists. I am not sure about that. Ethicists will anyhow need empirical premises if they want arguments concerning realistic problems to get off the ground. Should such premises be plugged in by scientists? My hunch is that many scientists will not regard *that* as their business.

Science and ethics are now separated by a wide gap. That is unfortunate because we often need to confront empirical issues and normative issues at the same time in attempts to solve practical problems.

Other areas bearing on science and values also manifest deplorable gaps. The last few decades have witnessed the development of a new discipline called technology assessment (TA), which aims to realize what its name expresses. In TA, theories of natural science and medicine with a bearing on technology play an important role. In addition to this, politics, social science and decision theory are considered.

Ethical theory is mostly disregarded in TA, although TA is clearly a normative discipline. In TA theories of social science are critically appraised. Such theories are important in view of decision processes, but they are notoriously controversial. Hence the need to appraise them. In contrast to this, theories of natural science and medicine are very much taken for granted. I hope that the previous chapters have shown that an appraisal of these theories is not superfluous. In my opinion TA needs to be expanded with methodological criticism of science now taken for granted.

The philosophy of medicine, unlike the philosophy of biology, *is not concerned with science in a concrete way*. Medical ethicists, for example, primarily deal with subjects such as patient rights, euthanasia, artificial insemination, and so forth. So they should. In addition, they should address the bearing of medical science on practice. True, they do this in a general sort of way. There are many discussions on the limitations of biological approaches of health and disease. However, *medical ethicists seldom have a close look at the biology involved*, to their detriment I would say (Van der Steen and Thung, 1988; Van der Steen, 1991, 1993b).

The examples below illustrate the crucial role which empirical issues should play in normative settings.

• *Example 1.* Medicine nowadays is faced with a serious problem of cost containment. Medical treatments have proliferated and medical technology is becoming increasingly expensive. Cost-benefit analyses are high on the health policy agenda. Ethics plays an important role in all this.

If full-scale application of an expensive treatment is not possible for lack of funds we need criteria to select patients who will get the treatment. Expected benefits in terms of the length and quality of life patients are likely to have after treatment are an example of a criterion.

In discussions about this, little attention is paid to the possibility that medical treatments may have no effect, or even be harmful. This is a serious oversight. Methodological considerations in *6.4* show that medical treatments should not simply be taken for granted. It is conceivable that extensive methodological research on treatment effects would lead to the conclusion that many medical treatments which are now applied as a matter of course had better be abandoned. Those who aim at cost containment should first and foremost consider this possibility. •

• *Example 2*. Egoism and altruism are important concepts in ethics. Ethicists do not always realize how complex they are (cf. *2.4*, example 2). This is a reason for criticism I will not elaborate. At least as important is the fact that ethicists almost totally ignore publications by researchers in empirical science on egoism and altruism. Perhaps biological research on this is not crucial for ethics, because the concepts of egoism and altruism have different meanings in biology. Psychology and social science, however, are clearly relevant.

Those who defend a doctrine called ethical egoism hold that selfishness is the ultimate justificatory base of morality. It seems to me that such a view will not have much practical interest, when realistic empirical issues concerning selfishness are disregarded.

I have read all recent publications on ethical egoism I managed to lay my hands on. In all of them reference to research by biologists, psychologists and social scientists is totally lacking. To the extent that the thesis that humans are basically selfish means something (it is actually unclear), I would regard it as false. Most humans are selfish part of the time, in some respects. To arrive at a sensible ethical theory we will have to be specific about this. Armchair theorizing will not easily provide the required information. Empirical science is a better source.

It seems to me that ethics is hardly in a position to formulate guidelines for human conduct as long as it disregards empirical research. •

• *Example 3*. Possibilities for artificial genetic modification of organisms are rapidly increasing. One aim of research on this is crop improvement, or the improvement of soils, say by the use of manipulated bacteria. Under what conditions should the release of genetically modified organisms (GMO's) in the environment be allowed? The issue is now controversial. Biotechnologists are relatively optimistic about risks, various ecologists are critical. In some cases researchers aim to keep GMO's confined to restricted areas, but they might escape. So it is important to know the probability that escapees will wreak havoc in existing ecosystems.

Optimism of biotechnologists concerning risks is often based on the following line of reasoning (see, for example, articles concerning bacteria in Fry and Day, 1990). GMO's are similar to spontaneously occurring mutants. Now mutants almost always have a decreased fitness. Moreover, in natural environments mutants without a decreased fitness seldom cause disasters. Therefore the use of GMO's in the field will not normally be risky. Any risks there are can be decreased by providing GMO's with genes that decrease fitness.

To evaluate this argument we need empirical evidence. In addition to this, methodological analysis is necessary. An important point is that 'fitness' is a many-place-predicate (see *2.5*). Organisms do not have fitness values *simpliciter*. Fitness is relative to the environment. It is always possible that a GMO which has a low fitness in environments considered by researchers will prove to have a high fitness in different environments. Highly general statements about fitnesses are therefore impossible (cf. comments on the theme of natural history in chapter 7). Thus there is no warrant for *general* optimism concerning risks.

Methodological analysis shows that we should be very cautious with GMO's in the field because an accurate determination of risk values is impossible. •

9.5. Bias: Miscellaneous Examples*

Any view we care to develop, in science or elsewhere, will necessarily be biased in the innocuous sense of selective. That is why the term bias is seldom used in this sense. There is also a *pejorative sense*, which I would describe as follows. A view, or an account, is biased in this sense if, in addition to being selective, it has the emphasis in the wrong place, misleadingly so. Important in bias are things left unsaid. In the evaluation of theories and arguments concerning relations between empirical and normative matters, bias is extremely important.

The examples below, which bring the textbook to its conclusion, illustrate that bias is widespread, and that it can take many different, subtle forms. The theme of bias is also prominent at the background of chapter *10*, which does not belong to the main text.

• *Example 1.* The concept of genetic determination is tricky (see *2.5*, example 2). All features of organisms, man included, are affected by genetic *and* environmental factors. Therefore the concept does not apply to features as such but rather to differences in features between organisms. Some features (for example the presence of particular diseases) are 'genetically determined' in the strong sense that differences between persons with the feature and persons lacking it are *always* due to a genetic difference. The vast majority of features, however, are not so determined. Loose talk about genetically determined diseases should therefore be avoided.

At present there is an international research program that aims at the charting of the whole human genome. The costs run into billions of dollars. One of the motives is the assumption that knowing the human genome will help us understand the genetic causes of diseases, and thereby find new treatments. If we know which genes are involved, and know how they affect human physiology, we could counteract adverse effects by manipulating physiological processes. That is what some publications would have us believe.

---

* In view of the general nature of this section there will be no exercises.

A full-scale methodological analysis of the crucial notions involved here should make us cautious. Some defenses of the program presuppose that strong genetic determination is relatively common. It is not. To the extent that there is no strong genetic determination, there is no reason to put genetic factors at the top of the research agenda (for further comments, see Van der Steen, 1993b). •

• *Example 2.* In some areas of medicine, most notably psychiatry, two main categories of etiological factors are distinguished, biological and psychosocial ones (see 2.5, example 3). The category of biological factors appears to stand for internal factors, external factors are taken to be psychosocial ones.

Two dichotomous classifications are obviously amalgamated here. The result is a classification which does not exhaust the possibilities. Factors belonging to the physical or the biotic environment could well influence psychiatric disorders. For example, it has been shown by some researchers that articles of diet are sometimes implicated. Such possibilities are disregarded in most psychiatry texts. At the present time no one knows how many people are confined to psychiatric institutions due to the fact that the food they eat does not agree with them.

Methodological analysis of concepts indicates that theories in psychiatry are biased. As a result, some patients will probably receive the wrong treatment. I think it would be wise to incorporate methodological criticism of theories in any program designed to improve treatments for psychiatric patients (see also the analysis of recent literature in Van der Steen, 1993b). •

• *Example 3.* The following quotation is from an editorial comment in New Scientist (2 May 1992).

> Population is undoubtedly one of the great issues of our times. But is it really driving us to the ecological abyss? Consider that 20 years ago we were told, notably by Paul Ehrlich ... that rising populations would cause a billion or more people to die from hunger during the 1980s when world food stocks were calculated to run out.
>
> What happened? There were famines, but science and human endeavour ensured that the world remained awash with food. The famines resulted not from a world food shortage but from a failure either to help peasant farmers in drought-prone lands or to distribute the world's spare food when crops failed. The failure was therefore economical and political, rather than in any way demographic (p. 3).

The analysis in this example aims to locate causes of a particular phenomenon, local famine. Two causes are identified: failure to help farmers, and failure to distribute the world's spare food. The idea is that the famine would not have occurred if we would have done certain things. The passage is misleading because it concentrates on certain causal factors and disregards others. One could also argue that the famine would not have occurred if the local population density would have been lower. Population density should be regarded as a causal factor if the others are looked at in this way. Thus the suggestion that demographic factors played no role is highly problematic. •

The Issue of Generality in Ethics*

9.A. Introduction

Generality has long been regarded as one of the most important criteria which scientific theories must satisfy. Likewise for philosophical theories. Awareness is increasing that possibilities for general theories in science are limited. For example, biology[1] and medicine,[2] unlike physics, have few general theories and laws. Their subject matter can be characterized to a large extent as 'natural history', that is, as non-general 'theory'. Indeed, philosophers of science today are unlikely to promote the ideals of pervasive generality *for science* which Logical Positivism fostered in the past. However, many of them endorse generality as an ideal for philosophy itself.[3] Why philosophy should be different from science in this respect is unclear to us. Concerning generality we would endorse an ideal of pervasive modesty.[4]

Within philosophy, ethics has a separate position since its representatives do not always accept the ideal of generality as a matter of course. To the contrary, many researchers nowadays argue that general theories are of no use in ethics. Some even argue that ethics cannot have theories at all.[5]

Criticism of generality in ethics takes many forms. We will pay special attention to the view that casuistry rather than general theory is the proper basis of ethical reasoning. Unlike some ethicists who give casuistry a central position, we do not regard lack of generality as a feature that distinguishes ethics from science. The subject matter of science does not simply consist of general theory.

We argue that the distinction of general theories and casuistry has been misconstrued since 'generality' is a composite notion. To deal with the distinction we need methodological tools which do not get the attention they deserve in ethics. We outline an elementary methodology, which partly agrees with R.M. Hare's views.[6] Hare is a typical proponent of abstract ethical theorizing which casuistry opposes. Casuistry has been re-defended in a recent book by Albert Jonsen and Stephen Toulmin.[7] By an appraisal of their views in terms of our methodology we can indicate why their approach need not be at odds with attempts to develop 'general' theories.

---

* Slightly modified version of W.J. van der Steen and A.W. Musschenga (The Journal of Value Inquiry 26: 511-524, 1992), reprinted by permission of Kluwer Academic Publishers. For literature see notes. Not covered by index.

Our methodology can be fruitful in science and in ethics alike. We hope that it will help to put correspondences and differences between science and ethics in a better perspective.

9.B. Generality from a Methodological Perspective

The term 'generality' has been a source of much confusion since it is used for various concepts. First, it may stand for *universality* of form. A statement is universal if it contains a universal quantifier and does not mention particular individuals, times, or places. Second, *generality* is the opposite of specificity. We will not elaborate a full-fledged definition of this notion. The following example will suffice to show that we need to distinguish it from generality in the sense of universality. The statement that organisms are killed under extreme environmental conditions is more general than the statement that they are killed under extreme temperatures. With respect to universality, the two statements are on equal footing. Third, the term 'generality' may stand for general *validity* of statements. The statement that all organisms contain carbon is generally valid, for it is true of all organisms (as far as we know). The statement that all philosophers are wise has a restricted validity. Without qualification it is false.

To avoid confusion, we will use the term 'generality' only in the second sense, for the opposite of specificity. The notions of universality and generality can be used to characterize both descriptive discourse (as in empirical science) and normative discourse (as in normative ethics). In the context of ethics, Hare has pointed out that universality must not be confounded with generality.[8] As an example, he mentions the two principles 'Never kill people' and 'Never kill people except in self-defence or in cases of adultery or judicial execution'. Both principles are equally universal, but the first is more general than the second.

The way we informally defined 'validity' is appropriate for the characterization of descriptive discourse. For normative discourse we need a different notion of validity if we are unwilling to apply the predicates of truth and falsity to normative statements. We assume that any specific meaning of 'validity' in the context of ethics is sufficiently clear for ordinary discourse. In Hare's terminology, (general) validity in ethics is a matter of being universally binding.

The notion of validity bearing on statements must be distinguished from another notion which applies to arguments; unfortunately, the same word is used for both notions. The validity of arguments is a matter of relationships among the statements they contain. Specifically, a deductive argument is said to be valid, or formally correct, if the conclusion cannot be false while the premisses are true. In the present essay the term 'validity', if used without qualification, always stands for statement validity.

The search for 'general' theories, in science and in ethics, has been concerned with universality, generality, and validity at the same time. In the philosophy of science, the three species of 'generality' have been regarded as methodological criteria which scientific theories must jointly satisfy. Concerning science and ethics alike, we argue that, in most cases, the criteria cannot be satisfied at the same time. Before that argument we need to introduce additional distinctions.

In principle, we can easily evaluate theories with respect to universality and generality. We only need straightforward logic (and linguistic analysis) for this purpose. With respect to validity the situation is more complex. In order to know that a statement is valid we need to justify it. Therefore, the criterion of validity must be coupled to matters of justification.

In science, the criterion of validity is embodied in the principle that theoretical statements (such as laws, theories) need to be well-confirmed. This principle expresses the demand that attributions of validity must be justified. Two major kinds of justification exist in science. First, we can justify a statement by testing implications against data. Deriving implications is a matter of logic, though not a simple one. Additional assumptions are always needed for inferences. Second, justification occurs in the form of coherence. Statements which cohere with a body of theory have a stronger position than isolated ones. Coherence relations are again a matter of logic. We will use the expressions *data-justification* and *coherence-justification* as a convenient shorthand for the two kinds of justification. Obviously, we will normally combine the two forms of justification in evaluating statements of science. Justification involves inferential relationships among statements (compare the notion of argument validity).

In the past, philosophers of science have defended strong versions of either form of justification. Data-justification was associated with the ideal of verification, while coherence-justification was associated with the ideal of neat, deductively organized theories. Nowadays both ideals are regarded as obsolete. With respect to justification, we need much modesty in science.

Validity (of statements) in ethics will differ from validity in science. Nonetheless, the distinction of data-justification and coherence-justification is as useful in ethics as it is in science. 'Data' in ethics will be covered by specific normative statements that we accept *and* facts which are relevant in normative contexts. Justification in ethics, as in science, will largely be a matter of logic. That is, we will typically argue from accepted data and/or generalities, and the arguments will have to meet criteria of logic such as formal correctness.

We may need deontic logic besides ordinary extensional logic in ethics to account for normative statements in arguments. However, this does not warrant the conclusion that the logic of ethics is very different from that of science. Evaluation in science calls for the application of normative methodology, so extensional logic will not suffice there either.

Consider the thesis that, barring exceptions, the use of persistent pesticides is morally wrong. We would justify this statement by inferring it from the following higher-level statements: Persistent pesticides accumulate in food chains. This results in forms of environmental damage we cannot accept for moral reasons. Also, the process of accumulation represents unacceptable health hazards since food will be contaminated.

The inference represents coherence-justification, in the form of deduction, on the basis of factual and normative premises. The premises themselves need to be justified by appeal to data from biology, data in the form of moral intuitions concerning specific examples of damage, etc., and generalities we take for granted (the process of justification must stop somewhere to prevent infinite regress).

Some philosophers think that justification is impossible in ethics. We concentrate on two of the arguments to this effect.

First, it has been argued that justification of concrete moral rules by inferences from theories is unsatisfactory because theories are typically appealed to *after* we have decided which rule to adopt. This is taken to imply that theories do not have justificatory force. This argument is a blatant fallacy since it confounds temporal priority with logical priority. Henceforth we will refer to it as the *priority fallacy*.

A parallel with empirical science should explain why this is a fallacy. The justification of a moral rule by an argument with general principles in the premises is the analogue of a scientific explanation in which phenomena are connected with laws or other generalities. Phenomena we want to explain will mostly be known before explanatory premises are formulated. Science would become impossible if explanatory arguments were rejected for this reason.

Second, some thinkers oppose general theories because they do not permit the derivation of moral rules that we encounter in daily life. This argument calls for a digression on the concept of theory. Discussions about the impact of ethical theories become confused since 'theory' is an ambiguous notion. Philosophers of science have shown that the concept of (scientific) theory is tricky.[9] Unfortunately, their publications on the subject are totally disregarded in ethics.

We will not try to cover all the senses of 'theory' in ethics. The admittedly simplistic distinction of *formal* and *substantive* theories will suffice in the present context. Likewise for formal versus substantive principles; theories are here conceived as collections of interrelated principles (relatively general statements).

Formal principles are methodological criteria, such as the ones we have introduced, for the appraisal of substantive principles. In science the distinction is straightforward. In ethics the situation is more complex. Statements such as 'killing is wrong', in the terminology we adopt, are substantive principles of ethics. The thesis that such principles must satisfy criteria such as universality, or universalizability, is obviously in an entirely different category. It has a formal nature. Now ethics also has principles which are not easily placed in one of the two categories. For example, the basic tenets of consequentialism and of deontology can be construed as high-level substantive principles, but also as formal constraints for the substantive principles we envisaged. We have no preference for any particular reconstruction. Either reconstruction might be sensible depending on the purpose the theorist has with an 'over-arching' theory.

Those who oppose general theories in ethics[10] use the concept of theory in a narrow sense for highly abstract theories, especially formal ones. Such theories allegedly do not allow the derivation of 'moral rules' we use in daily life. Yet such rules can be regarded as substantive principles at intermediate levels of generality. The thesis that theory has no use in ethics is therefore misleading. It typically pre-supposes the adequacy of lower level generalities which are denied the status of theory for no obvious reason.

However, the critics of theory do have a point since theorizing in ethics often proceeds at an unduly abstract, formal level. We agree that substantive principles (such as 'moral rules') cannot be derived from abstract, formal principles alone. For example, any *formal* thesis to the effect that ethics must aim at a 'reflective equilibrium' does not imply much about substantive principles. Reflective equilibrium is a variant of coherence, a methodological criterion which constrains such principles but does not provide them with content.

Antitheorists are utterly mistaken if they conclude from this that ethics, unlike science, has no use for theories. In science we encounter the same situation. No scientist in her or his right mind would try to derive Newton's laws from method-ological precepts, for example the requirement that laws of nature must be univer-sal.

We will not analyze in more detail here disputes concerning the role of formal theories. Our aim is more modest. We want to provide tools for the improvement of concrete ethical discourse. We think, though, that problems with formal theories of ethics cannot be solved unless ethical discourse is clarified with mundane methodology.

9.C. Methodological Trade-Offs

In the sequel we will use the term 'ethical principles' for substantive ones. The va-lidity of ethical principles can be restricted in many different ways. We will use the term 'condition' (symbol: C) to cover the general idea of restriction. The term may stand for items as diverse as cultures, historical periods, localities, and situations actors may find themselves in. In the analysis presented below we assume that contexts which differ in conditions, however specified, allow of rational compari-son. That is, we will side-step problems associated with strong forms of rela-tivism.[11]

From a formal point of view, conditions which restrict the validity of principles can be accommodated in several ways. Consider the principle 'killing is wrong'. Suppose, for the sake of argument, that we agree that the principle is valid, with the restriction that conditions C_i represent exceptions. In that case we can express our opinion in two ways. (1) We say that we accept the principle, though, if confronted with an exception, we will be explicit about C_i overruling it. (2) Right from the start we formulate 'killing is wrong unless C_i' as our principle.

Notice that (1) and (2) are logically equivalent. The differences in formulation are nonetheless interesting. The principle in (1) is highly general though it is not (generally) valid. The principle in (2) valid because it accommodates the exceptions. Its validity is attained through a decrease in generality. This shows that methodological criteria can be at cross-purposes, so that trade-offs need to be faced. We are using the notion of trade-off for formal issues; substantive trade-offs we will have to face when principles conflict are an entirely different matter.

In the example, generality and validity cannot be maximized at the same time. So we are faced with a trade-off problem. Which criterion should we privilege? That is a question without a general answer. The context will determine which criterion is more important.

The example is an artificially simple one in which the issue of trade-offs is unexciting. In more realistic settings, however, trade-offs need to be faced which are by no means trivial. We should recognize that the search for a theory which satisfies sundry methodological criteria in a high degree is futile. Instead we should be content with theories which are adequate in some contexts, though inadequate in others, in view of the particular limitations they have. In science, this is nowadays a commonly accepted way of looking at theories. Ethics lags behind in this respect.

Exceptions to ethical principles are seldom available in a fully articulate form. We will often come across new situations we had not thought of before, which call for the recognition of new exceptions. That is what makes ethics difficult and exciting at the same time. In this respect, ethics does not differ at all from science or any other intellectual endeavor. Theories, and any principle they contain, are always open-ended. Thus, if we would aim at a really valid principle concerning killing, we had better give it the form 'killing is wrong except in conditions C_i', where C_i stands for conditions partly or wholly characterized in vague terms. That is what makes the principle open-ended.

While a principle of this kind would not be wholly clear and articulate, would this make it inappropriate? Being clear and articulate is another methodological criterion which principles ideally should satisfy, but here again we must be prepared to face trade-offs. We can opt for fully articulate principles (in this case by deleting C_i) which will turn out to have a restricted validity, or we can aim at general validity and pay the price of vagueness. The context will determine which option is better.

To illustrate the use of the methodological framework we have introduced, we will apply it to two theories which are quite different with respect to generality, that of Hare and that of Jonsen and Toulmin.

9.D. Hare's Views of Generality

Hare has defended the thesis that moral reasoning can take place at two levels, the intuitive one and the critical one.[12] At the intuitive level we are dealing with *prima facie* principles, at the critical level with critical principles. According to Hare, the function of *prima facie* principles is to help us reach decisions in concrete, practical situations, and they can only perform this function if they are simple. If conflicts occur, we must not try to improve on principles geared to the intuitive level by making them more complex, but turn instead to the critical level. Conflicts can only be solved by appeal to critical principles.

Hare maintains that both *prima facie* principles and critical principles are universal. The difference lies in the generality-specificity dimension. A *prima facie* principle has to be relatively simple and general (that is, unspecific), but a critical principle can be of unlimited specificity.

The following distinction concerning universality is important. Universality may be a matter of form, and it may involve issues which are universally binding. In our terminology, universality and validity should not be confounded. *Prima facie* principles are universal, but they need not be (generally) valid. Critical principles are both universal and valid.

Though we are critical of Hare's category of critical principles, his analytical tools are quite useful. Consider the following well-known problem. Normative ethics has been said not to deal adequately with conflicts in which we are naturally partial. Is not a mother entitled to love her child more than other children? Yet, ethics calls for an attitude of impartiality. Altruism and love must be extended to people in general, otherwise we will violate universal and general principles.

Hare dismisses this kind of reasoning—rightly so we think—as follows.[13] At the intuitive level, the example involves partiality. But critical thinking may show that a universal statement to the effect that *every* mother should love her own child more than other children may be valid. At the critical level, partiality need not occur in this case. Those who have dealt with such cases in ethics have often confounded the issues for lack of an articulate methodology.

All this fits in with the methodological distinctions introduced above. However, we are not happy with the way Hare develops his theory. Hare does not give articulated examples of critical principles. His main concern is with defining features of critical principles, namely universalizability, prescriptivity, and overridingness. Indeed, in an earlier work[14] he explicitly stated that moral principles need to satisfy criteria such as universality, but they need not be formulatable.

Hare does not explain why he refrained from articulating critical principles. A possible reason is that only superhuman creatures would be able to elaborate them. If that is what he thinks, we would not be interested in critical principles since we are human. Alternatively, the principles could be person-relative. If that is what Hare means, he rightly gives no examples since Hare-relative principles would not be binding on his readers. We would not be interested since we are not Hare. Yet another interpretation is possible, which may be plausible in view of Hare's thesis that critical principles, unlike *prima facie* principles, can be of unlimited specificity. In many cases if not always we may need very complex and specific statements to arrive at universality. Perhaps Hare did not feel the need to complete this task because he was satisfied with the conviction that it can be completed *in principle*.

The second interpretation, which would make critical principles person-relative, is supported by Hare's views of universalizability.[15] A moral judgment about some situation is universalizable in Hare's sense if *the person involved* is prepared to make the same judgment about any other (actual or hypothetical) situation which is precisely similar. Other situations include those in which the person occupies the position of another party; an exchange of roles should not make a difference.

The *precisely similar* enables Hare to bypass the tricky notion of morally relevant features. The situations Hare considers are qualitatively identical, so the question whether they are sufficiently similar to bear on universalizability does not arise.

This looks like a consistent position. However, Hare's stance is singularly unhelpful if we want to implement the criterion of universalizability in order to *formulate* a critical principle. To try to accommodate *all* the features of the situation concerned in a principle is absurd. (The notion of a set of 'all' features is incoherent.) So we are forced to work with features we judge to be relevant. The chances are that any particular situation will involve a great number of such features. Also, infinitely many kinds of situations exist. For these reasons, many authors have argued that Hare's universalization thesis is vacuous or trivial.[16]

We admit that the issue is elusive since Hare could insist that persons are able to envisage identical situations without bothering about specific features. We are unable to perform this kind of envisaging, but the failure may be ours. However, if specification of features is really dispensed with, we have a different objection. A critical principle we arrive at in this way would essentially be about a *particular* situation and situations which are similar to it. Moreover, person-relativity would imply that reference to a *particular* person is a covert part of the meaning of critical principles. On two counts, therefore, critical principles would not be universal! When Hare states that such principles are universal, he means that they contain a universal quantifier. That is not sufficient for statements to be universal in the normal sense of the term in logic and philosophy. Besides containing a universal quantifier, universal statements should meet the requirement that they do not mention particulars.

To some extent, this is a matter of terminology. Hare is free to use the term 'universal principles' for items which are not formulated and which essentially involve particulars. However, the chances are that such idiosyncratic conceptualizations will generate much confusion. We can but conclude that Hare's critical principles should not be regarded as components of an ethical theory (unless we *also* choose to use the concept of theory in an idiosyncratic way).

If critical principles in Hare's sense could be formulated, they would be universal (in some respects at least) and valid. In many cases we would have to pay a price for these desirable features, in the form of lack of generality and, perhaps, clarity. *Prima facie* principles are universal and general. Their disadvantage is that they are not generally valid. This classification of principles is not exhaustive. For example, in some contexts we could be content with principles which are general, valid, and non-universal.

Hare's treatment of critical principles is highly confusing. Yet his work remains valuable because he has shown that ethical principles cannot satisfy all apparently reasonable methodological criteria. The context of interest will determine which criteria we should emphasize.

9.E. Comments on Casuistry: Jonsen and Toulmin

The most elaborate defense of casuistry is found in the recent book by Jonsen and Toulmin.[17] We will give a running commentary on their views to assess the merits of casuistry *vis-à-vis* overly abstract, general theories.

Jonsen and Toulmin notice that highly abstract generalizations in ethics are tailored only to fit paradigm cases. They are not suitable to resolve ambiguities and conflicts among principles. Jonsen and Toulmin present the following telling example.[18] The National Commission for the Protection of Human Subjects of Biomedical and Behavioral Research, which was set up in the United States in 1974, managed to reach agreement about many particular moral judgments. But the moment discussions soared to the level of abstract principles, no consensus could be reached.

> In theory ... particular concrete moral judgments should have been strengthened by being 'validly deduced' from universal abstract ethical principles. In practice the general truth and relevance of those universal principles turned out to be *less* certain than the soundness of the particular judgments for which they supposedly provided a 'deductive foundation'.[19]

We agree, but we think that Jonsen and Toulmin's terminology is misleading. They contrast abstract generalizations with particular moral judgments. The term 'particular' in philosophy often has the connotation of non-universality. We can speak of rocks in general and formulate universal statements about them and about particular rocks. This is doubtless not what Jonsen and Toulmin have in mind. We assume that their particular judgments are universal statements at a low level of generality. They are opposing ethical theories which put extremely general statements at center stage.

Jonsen and Toulmin subsequently contrast theory and practice. They characterize theory as idealized, atemporal, and necessary. By contrast, practice (such as ethical reasoning) is concrete, temporal, and presumptive. In this context the following remark about practical judgments is interesting.

> Once this practical judgment is exercised, the resulting decisions will (no doubt) be 'formally entailed by' the relevant generalizations, but that connection throws no light on the grounds by which the decisions are arrived at, or on the considerations that tilt the scale toward one general course of action rather than the other. What such decisions involve can be explained only in *substantive* and *circumstantial* terms.[20]

Again we agree, yet again we note that Jonsen and Toulmin's formulations are misleading. The phrase 'the grounds by which the decisions are arrived at' concerns the temporal priority of some statements in moral reasoning. 'Formally entailed by' stands for logical priority. Jonsen and Toulmin are committing the *fallacy of priority* (see section B). Their line of reasoning could even be rebutted by those who would like to set up ethical theory in the spirit of geometry. Mathematicians at times arrive at brilliant theories by capricious activities. They are welcome, as long as the results are good.

Jonsen and Toulmin's key terms 'substantive and circumstantial' (which keep cropping up in their book), likewise, have less force than they suggest. Sound ethical principles, at any level of generality we may wish to consider, will need to have substance in order to be useful. In applying them in actual practice, we will have to specify 'circumstances'. Proponents of abstract approaches in ethics should admit that much. Consider the following analogue with scientific explanation. Newton's laws will permit us to explain that a stone falling from the window will hit the ground at a *particular* moment. However, we cannot infer this from the laws alone. We will also need initial conditions such as the moment the stone started to fall. This is a 'concrete, temporal [though not presumptive]' matter, described in 'substantive, circumstantial' terms.

The distinction of theory and practice is not as sharp today as it was in antiquity. But essential differences prevail.

> Another feature of the analytic contrast between Theory and Practice concerns the solidity of argument in each. Within scientific theories today arguments are no longer accepted on a priori grounds alone, but they are still 'necessary' in a less ambitious sense. So long as any scientific conclusion follows from theoretical principles strictly, that inference is valid formally as much as substantively. Conversely, when practical arguments go beyond the scope of any formal theory their conclusions are 'presumptive' in a similar sense. Their soundness depends not on formal validity alone but on the richness of the substantive support for any general ideas they use and the accuracy with which any particular case has been recognized and classified.[21]

Jonsen and Toulmin subsequently note that inductive reasoning and pattern recognition are extremely important in practical arguments. In theoretical arguments the emphasis is more on deduction.

We think that the authors are again exaggerating the differences between theory and practice. In science as well, researchers often 'go beyond the scope of formal theory'. Inductive reasoning is then legitimate. And presumptiveness is often regarded as a hallmark of science.

Jonsen and Toulmin subsequently present an impressive survey of the history of casuistry. They deplore the decline of casuistry in the last few centuries, and they suggest that we can learn much from the ways of the old casuists. In the concluding chapters the case for casuistry is strengthened on the basis of this. The following quotation captures the essentials of the main conclusions.

> Every well-founded ethical theory carries conviction on some occasions, in some circumstances, applied to problems of some kinds; but no theory has a monopoly in all situations or over all kinds of problems. ... The objections to thinking of ethics as a 'science' ... are as strong as ever. ... What patterns of argument are appropriate in dealing with any particular kind of problem must ... be judged *contextually* with an eye to the specific case at issue.[22]

This must not be taken to imply that ethics cannot develop theories at all. The point is that it must not be modelled on the exact sciences. For parallels we have to look elsewhere in science.

We argue that the way Jonsen and Toulmin characterize ethics should fit some exact sciences quite well. As we noted, biology and medicine do not have many general theories. Jonsen and Toulmin's point is that ethical theories which are general and universal are unlikely to be valid under all conditions. We agree, but we would add that conditions of applicability can be accommodated by principles of a theory. Thus we could buy validity at the cost of generality, while retaining universality. If we cannot specify all the relevant conditions, we will need to be content with open-ended principles.

On this interpretation, what Jonsen and Toulmin are saying is fully compatible with the methodology we have introduced and, to a large extent, with Hare's methodology. (Hare is an obvious example of ethicists with a preference for the abstract theories Jonsen and Toulmin are opposing.) Unfortunately, Jonsen and Toulmin themselves are not sufficiently aware of the methodological distinctions we have discussed.

Jonsen and Toulmin do state that their view is apparently compatible with the more traditional abstract approaches. Their claim that casuistry is unavoidable in the application of ethical theory, will not be disputed by moral theologians or philosophers with abstract tastes and theoretical inclinations. However, the authors also purport to defend a less trivial claim, "that *moral knowledge is essentially particular*, so that sound resolutions of moral problems must always be rooted in a concrete understanding of specific cases and circumstances".[23] That is, in moral reasoning we should follow a bottom-up approach rather than a top-down approach.

All our previous comments are applicable here. First, 'particular' in the intended sense is not an opposite of 'universal'. It refers instead to generality at low levels. The same goes for 'specific cases and circumstances'. Once this is recognized, we can grant that much moral knowledge, to the extent that consensus exists, is essentially particular. So is much knowledge in the natural sciences. Second, the *priority fallacy* lurks beneath the surface. Jonsen and Toulmin again do not distinguish temporal and logical priority. Maybe bottom-up approaches are quite suitable to arrive at moral knowledge (temporal priority) whereas a reconstruction of this knowledge is more profitably cast into top-down form (logical priority).

9.F. Conclusion

Does ethics have adequate general theories? Our analysis shows that this question does not have a straightforward answer since the key terms are ambiguous. So we should not concentrate on the answer but on the question itself. 'Ethics' stands for many things, but we let that pass. 'Adequate' may refer to varied arrays of methodological principles which are seldom fully articulated in ethics. 'General' is a notion with at least three meanings. Different kinds of generality may be at cross-purposes, so we must not expect theories to be general in sundry senses. 'Theory', for that matter, is itself ambiguous. Some thinkers say that ethics cannot have theories, while others deny it. We doubt whether opposing parties are talking about the same things.

No wonder then that controversies in ethics are long-lasting and unproductive. We hope that the methodology we have presented will alleviate some of them. The examples we chose show that this is feasible. Views such as Hare's and Jonsen and Toulmin's, which are seemingly wide apart, show convergence if we put them in a methodological perspective.

Our analysis also suggests that many alleged differences between science and ethics could fade away if methodology is brought to bear on them. Specifically, the idea that ethics compares poorly with science in view of limited generality, or poor means of justification, is unfounded. Those who defend this view over-rate the powers of science.

Notes

1. Wim J. van der Steen, "Concepts of Biology: A Survey of Methodological Principles", *Journal of Theoretical Biology* 143 (1990): 383-403; Wim J. van der Steen and Harmke Kamminga, "Laws and Natural History in Biology", *British Journal for the Philosophy of Science* (in press). References concerning the views of several philosophers of biology are given in the latter article.

2. Kenneth F. Schaffner, "Exemplar Reasoning about Biological Models and Diseases: A Relation between the Philosophy of Medicine and Philosophy of Science", *Journal of Medicine and Philosophy* 11 (1986): 63-80; Wim J. van der Steen and Paul J. Thung, *Faces of Medicine: A Philosophical Study* (Dordrecht: Kluwer, 1988).

3. Some philosophers, however, have explicitly stated that the search for generality in philosophy itself is mistaken. See especially Richard W. Miller, *Fact and Method, Explanation, Confirmation, and Reality in the Natural and the Social Sciences* (Princeton: Princeton University Press, 1987).

4. Peter B. Sloep and Wim J. van der Steen, "A Natural Alliance of Teaching and Philosophy of Science", *Educational Philosophy and Theory* 20 (1988): 24-32.

5. See articles in Stanley C. Clarke and Evan Simpson (eds.), *Anti-Theory in Ethics and Moral Conservatism* (Albany, N.Y.: SUNY Press, 1989).

6. R.M. Hare, *Freedom and Reason* (Oxford: Oxford University Press, 1963); R.M. Hare, *Moral Thinking: Its Levels, Method, and Point* (Oxford: Oxford University Press, 1981).

7. Albert R. Jonsen and Stephen Toulmin, *The Abuse of Casuistry: A History of Moral Reasoning* (Berkeley, Cal.: University of California Press, 1988).

8. Hare, *Moral Thinking*, pp. 40-41.

9. In the philosophy of science, two opposing views of scientific theories, the received view and the semantic view, are widely discussed. See e.g. Peter B. Sloep and Wim J. van der Steen, "The Nature of Evolutionary Theory: The

Semantic Challenge", *Biology and Philosophy* 2 (1987): 1-15. Other sources are mentioned in the article.

10. Clarke and Simpson, *Anti-Theory*.

11. For discussions of forms of relativism, see e.g. Michael Krausz (ed.), *Relativism, Interpretation, and Confrontation* (Notre Dame, Ind.: University of Notre Dame Press, 1989).

12. Hare, *Moral Thinking*.

13. Ibid., pp. 137-140. For analogous examples, see Bert Musschenga, *Mag het Hemd Nader Zijn dan de Rok? Over Grenzen aan de Morele Eis van Onpartijdigheid* (Amsterdam, Free University Press, 1989).

14. Hare, *Freedom and Reason*, p. 47.

15. Hare, *Moral Thinking*, pp. 42-43. We are only discussing universalizability in one sense of the term. For a discussion of varied meanings of 'universalizability', see A.W. Musschenga, *Noodzakelijkheid en Mogelijkheid van Moraal* (Assen: Van Gorcum, 1980).

16. See, e.g., G.F. Gaus, *Value and Justification, the Foundations of Liberal Theory* (Cambridge, England, Cambridge University Press, 1990), pp. 306-315.

17. Jonsen and Toulmin, *Abuse of Casuistry*.

18. Ibid., pp. 16-20.

19. Ibid., pp. 18-19.

20. Ibid., p. 30.

21. Ibid., p. 32.

22. Ibid., p. 297.

23. Ibid., p. 330.

CHAPTER 10

Disciplines and Dutch Dikes*

In all disciplines we desperately need generalists who cherish common sense and are aware of context-dependence.

Specialists can't oversee their own discipline any more. A few years ago some colleagues told me they attended a congress, in neurobiology I think it was, with many parallel sessions. Number of papers delivered: 20,000. If you visit a library and skip through issues of *Current Contents*, a fat *weekly* (actually a combination of several weeklies) with titles of new scientific articles, you will also get the picture I am trying to evoke. Maybe you already know about *Current Contents*. If so, then you will be familiar with the current staggering growth in numbers of disciplines and interdisciplines.

Do we need all this new research? I am skeptical, but I won't go into that.

Science has become unbalanced. Ours is the era of specialists. We desperately need generalists. Reviewing what we have, that should be a prime focus. It could help us uncover shortcomings: researchers writing down what was known elsewhere decades ago, contradictions on a massive scale, serious gaps and oversights, and so forth.

Can one be a generalist nowadays? I try to be one, and I do know it's a hard job. Science has developed a plethora of languages. Even the most gifted researchers can but master a tiny fraction of them. Thus the bad scene of current science is allowed to perpetuate itself.

---

* The informal style which is used in this chapter isn't compatible with extensive documentation. Hence the absence of reference to the literature and research published elsewhere. I am trying to evoke a point concerning biased trends in extant research. The visible bias in my own presentation aims to counteract these trends. The emphasis in this chapter is on a case study involving biology. For a similar study concerning medicine see Van der Steen (1993b). Implications for education are described in Van der Steen and Sloep (in press). A stimulating discussion with Ken Schaffner is the ultimate source of this chapter. I owe the final version to incisive comments from Annelies Stolp. The responsibility for the result is entirely mine.

I can only see one solution. We should aim to change the languages of science. Technical they are. So they must, *up to a point*. I have read many science texts with a message obscured by specialist jargon. Messages I thought I understood mostly gave me the feeling that we need more ordinary language and common sense in science. At times I could not find any message at all. The shortcoming may be mine.

I can't describe any new breed of generalists. That's something for the future. The rest of my story is a modest attempt to deal with prerequisites, ordinary language and common sense fed by an awareness of context-dependence.

Let me proceed with a free translation of a front page article from yesterday's issue of one of the best newspapers in The Netherlands, the NRC. Today is July 23, 1992. I won't deal with developments which occurred after today.

Lawsuit Against River Dike Reinforcement

The alliance *De Ooijse Dijken* [the dikes around the polder called Ooij; polder = extended flat pasture below level of surrounding water - WS] will apply for an injunction against the polder authority *Maas en Waal* [two rivers - WS] to postpone dike reinforcement along the polder Ooij where the Waal flows east of Nijmegen [city - WS]. The polder authority wants a building contractor to start reinforcing a five kilometer stretch of dike within a couple of weeks. This would mean the destruction of a precious river landscape.

Realization of the plans would result in massive tree felling on either side of the dike, demolition of houses, the filling up of a Waal tributary, and a five foot increase of dike height. [Note. The dike is a so-called winter dike against flooding; locally there is a lower summer dike closer to the river. - WS]

The attorney for the alliance will plead that damage to be inflicted calls for postponement because the *Tweede Kamer* [the parliament - WS] is about to reconsider standards for dike reinforcement recently elaborated by *Rijkswaterstaat* [Department of Public Works, Ministry of Transport and Public Works - WS].

The attitude of the *Tweede Kamer* resulted from a *D'66* [political party - WS] initiative aiming at a majority vote to reconsider these standards.

The attorney will demand that the polder authority be forced not to grant an assignment to any contractor for the time being or, in case an assignment has been granted, that the contractor be forced not to implement plans.

The secretary of the polder authority has stated that granting the assignment will take another month. He says that the polder authority refuses to consider postponement because it would greatly increase costs which already amount to

12,000,000 guilders.

As it happens *Rijkswaterstaat* is stating now through a spokesman that its own standards are criticized by officials within the Department. The critics argue that continuing protests against dike reinforcement, which will impoverish landscape throughout the river area, indicate a pervasive lack of social support. Hence execution of current plans appears to be problematic.

The rest of the article, on p. 2 of the newspaper, is summarized below.

The critics within *Rijkswaterstaat* do not think that the standards are wrong. They merely want to increase social support through an independent committee. Likewise for polder authorities in the river area. In financial terms the stakes are high. According to them, costs of dike reinforcement should largely be drawn from a national budget. Without reinforcement local authorities would be responsible financially for dike repairs long overdue.

What disciplines could help us get the issue in perspective? Let me give you some pointers.

The science of politics. Which authorities are responsible? How is power distributed? Which procedure is acceptable? Who is to pay?
Economy. Which are the costs and benefits, in the short term, in the long term? Do cost-benefit calculations favor a particular option?
Sociology. How do authorities exercise responsibility? How is public opinion shaped? How can those affected gain influence?
Hydrology and climatology. What are the chances of floods? What consequences does this have for dike reinforcement?
History. What is the origin of the present landscape? In what senses is it special? Is there a historical basis for preservation?
Biology. Is the area a precious one in view of animal or plant diversities? Does it have special plant or animal communities?
Ethics. How should the interests of all parties be weighed? What responsibilities do we have towards the persons involved? What about moral obligations towards animals, plants, biological communities in the area?
Philosophy. What's the value of theories in all these disciplines? Could we integrate them so that we get a unified picture?

Feeling uneasy? Well, I am. My list is by no means exhaustive. I won't continue in this vein. It's hardly possible to visualize the pile of knowledge from science and other disciplines that could be brought to bear on the issue.

Philosophy should help us sort out contributions from other disciplines but it doesn't. It has failed to take a really practical turn. That is a message I have tried to convey in this book.

Let us continue with *ethics*. In all the issues concerning the Ooij polder moral items abound. Could ethicists help us? My impression is that they are not equipped to do this just now.

More than ninety percent of ethics wholly disregards animals and plants. That alone would disqualify ethics to deal with the Ooij polder.

Let me portray UE, the Ultimate Ethicist. There are ethicists whose writings suggest that they regard themselves as UE. That at least is an impression I get when I read highly general books in ethics that present the one theory that should be the basis of all. Various books of this kind have been written, some of them quite recently. UE does not really exist of course. You can easily prove this if you are willing to read some of the books I am alluding to. Here is the gist of the proof in succinct terms. The theories in different books are very different. Hence the one theory does not exist. Hence UE does not exist. End of proof.

Responses to books with the UE flavor have been violent even within ethics, not least in the last few years (see appendix to chapter 9). Opponents have charged that ethics should do without theories. I think that the controversy is spurious. Opposing parties use the term 'theory' and many other terms in different senses.

I don't know what the opposing parties would say about the Ooij polder.

The dispute I referred to is not representative of ethics as a whole. Ethics is an extensive field. Ethicists are concentrating on many important practical problems. Their choices of problems are selective. For that they should not be blamed.

I see one major problem with ethics. Ethical problems are everywhere. Should ethicists be everywhere? That's not realistic. I guess we need to rethink the balance between science (if ethics is that) and common sense.

Considering dike reinforcement I would say that the first thing to do is visiting a few polders in the river area. I've done that. Here is an appropriate observation which as it happens does not concern the Ooij, but an area nearby along the river Waal. I visited a restaurant located near the river against the winter dike, at a spot where the water flows near the dike. Talked with a few people there. It was very easy to spot recent maximum water heights by the observation of debris left by the water. The restaurant was in danger of being undermined. At different nearby locations the situation was entirely different because of vast stretches of polderland outside the dike. It seems to me that this observation could undermine the idea that dike reinforcement should be uniform over the entire area.

I grant that my interpretation may be erroneous; it results from a single Sunday afternoon walk. However, the message should be clear. We need a different balance of science and common sense. Specifically we should distrust generalities allegedly based on science.

Lack of expertise does not allow me to consider all the remaining disciplines. Allow me to put *biology* at centre stage. I've visited the Ooij-polder and looked at organisms one comes across in the area. Let's concentrate on one issue, the diversity of organisms. If you want to observe diversity around you, you will need some background knowledge. I have some background knowledge on which I will rely.

What about relative diversity in the area? Let's be careful with the question. Which organisms do you want to consider? The diversity of birds is very high in the area. Butterflies and plants are less diverse, but there are local exceptions. You may observe the same situation in many parts of The Netherlands.

I recently had a conversation with an official responsible for the preservation of nature in various districts. She told me about professors of biology jointly helping her with the selection of candidate sanctuaries. Separate advices she got made sense. As a whole they were contradictory. Much depended on the target organisms the professors were researching.

Concerning diversity we will need measures. Numbers of species are one possibility. Biologists have also devised more complex measures, for example, indices that take relative abundances into account. Which measures would be appropriate? Considering bird diversity in the Ooij polder my answer would be that it doesn't matter much. Whatever technical measure we use in this context, diversity will turn out to be high compared to that in many other areas in The Netherlands.

Notice that I just took a stance concerning another problem with diversity. We need standards of comparison. By way of an aside, let me work this out in a general way. We look at birds and we compare the situation in Western Europe with that in the US. Let's concentrate on areas at the same latitude. Surprise. Diversities are very different. The US is more diverse. What would this mean? Partial answer: mountain ranges in the US often run from the North to the South, in Europe the direction is mostly from the West to the East. In the ice ages, US bird species could more easily shift ranges and so survive than in Europe. This is a historical explanation which uncovers a particular type of context-dependence in biology. General theories about diversity must be treated with caution. Their validity may be restricted to local situations.

One can also look at diversity by way of communities. Take plant communities. In the Ooij polder there are some sites with rare communities. How should we evaluate them? When I visited them I got the impression that they are special. In the case at hand I would tend to rely on this impression though it is merely based on amateurish field knowledge. Let's move on to science, though. Community ecologists in Europe have long ago divised a system of classification for plant 'associations'. It is still used in some quarters, but it is controversial. Some decades ago, researchers in the US developed a very different approach called ordination. The idea behind it is that plant communities in the form of neat associations do not exist. Do they? I won't discuss all sorts of details. The point of primary importance is that the US and Europe, in this case as well, are very different biologically. In the US gradual transitions are much more common. Again we are dealing with context-dependence concerning the validity of scientific knowledge.

I am convinced that the Ooij polder is special from a biological point of view. That is a point in favor of preserving it in its present state. My conviction is inspired by biology. At the same time I realize that biology cannot offer much in the way of generalities.

Beware of generalities preached in the name of science.

Answers to Exercises*

2.1.1. Your reaction should be that the disputants are using the term 'nervous system' in different ways. Their disagreement is due to conceptual confusion.

2.1.2. A is an empirical statement, *B* is a logical statement.

2.1.3. GAIA is an organism (if it shows self-regulation) if 'organism' is given an unusually broad meaning. The thesis that the earth shows self-regulation is an interesting empirical statement. The possible implication that the earth (GAIA) is an organism is a matter of terminological convention.

2.1.4. The first and the last sentence are logical statements that are part of the definition of 'vitamin'. The middle part of the passage is empirical.

2.1.5. "What connects these conditions ... to well-being ... is that they are some threat ... . Being a threat ... is a necessary condition for ... concepts." In the second sentence Goossens proposes that 'connects' in the first sentence stands for a logical connection. He gives no reason for this. The connection could also be regarded as an empirical one. Therefore Goossens defense of normativism is inadequate.

2.2.1. a. Too broad, mentions accompanying features. b. Not sufficiently clear, too narrow (cf. unicellular organisms). c. Too narrow (cf. asexual reproduction).

2.2.2. You should not allow this because the definitions are circular.

2.2.3. a. No. b. Yes. c. At the very least we should add a phrase at both sides which specifies the time needed to reach 50% mortality. Furthermore we should realize that the experimental set-up and the condition of animals must be taken into account.

---

* Not covered by index.

2.2.4. a. No. Factual evidence cannot compel us to change the meaning of words. One can decide, though, to change terminologies in the light of evidence if existing concepts turn out to be inappropriate, for example, because they have no empirical reference. In the present case there is no reason to change terminology. b. Notice first that the definition given is ambiguous. Is a chemical supposed to be a fungicide if it kills *all* fungi at *all* concentrations above some threshold? Or is it a fungicide if it kills *some* fungi at *some* concentrations? On the first interpretation, some of the chemicals are not fungicides. On the second interpretation there is no warrant for this conclusion.

2.2.5. The thesis that evolution without genetic change is impossible would be a logical statement. Therefore *no* evidence could lead to its acceptance or rejection.

2.3.1. The units of transmission, mutation and recombination are different. Therefore the concept as defined would have no empirical reference.

2.3.2. No, oxygen consumption would be an accompanying feature of 'energy consumption' which can be used as a meaning criterion. Being a meaning criterion does not make it a defining feature.

2.4.1. The conclusion of the authors does not follow from their theses. The reason is that their classification of limiting factors is not exhaustive. Limiting physical factors need not take the form of meteorological catastrophes.

2.4.2. The falsity of phyletic gradualism would not in itself entail the truth of punctuated equilibria, because the two models do not exhaust the possibilities. For example, it is conceivable that new species develop rapidly, and that the transformation involves large numbers.

2.5.1. No, the concept would have no empirical reference.

2.5.2. Mikhail aims at an integrative empirical theory. However, he forges a logical connection (cf. the definition) between the items which should be linked by the theory. This confusion of logical and empirical matters would make the theory inadequate.

2.5.3. Gould and Lewontin's formulations do not reveal the complexity of the concepts they define. As a result of this, the passage quoted misleadingly suggests that a particular form of adaptation can be ascribed to a particular feature (state, process). In reality a feature can show different forms of adaptation under different comparisons.

I will elaborate this for physiological adaptation and genetic adaptation, the first and the third form of adaptation mentioned by Gould and Lewontin. The following simplistic definitions seem to cover the intentions of the authors.

'Physiological adaptation' $=_{df}$ 'good fit, etc., environmentally determined'.

'Genetic adaptation' $=_{df}$ 'good fit, etc., genetically determined'.

(I am taking the good fit for granted; some would argue that there are also failed adaptations, which are adaptations none the less, in which the fit is not good.) Under both definitions, we will need to specify an environment relative to which there is adaptation, and we will need a criterion of 'goodness'. An even more important point is that any design or feature is 'determined' by environmental *and* genetic factors. Therefore the concepts of environmental determination and genetic determination in the definitions must not be applied to features (or designs) of organisms, but to differences in features between organisms. This implies that a feature of an organism cannot, as such, represent a physiological adaptation *or* a genetic adaptation.

Consider the following hypothetical example. Some organism O_1 developing in environment E_1 has design D_1 in this environment; in E_2 it would develop D_2. Suppose that D_1 satisfies a criterion of 'goodness' in E_1 (not in E_2), whereas D_2 satisfies the criterion in E_2 (not in E_1). The comparison in this description justifies the statement that O_1 shows *physiological adaptation* to E_1. Suppose that another organism, O_2, develops D_2 in E_1. If O_1 in E_1 is now compared with O_2 in E_1 (rather than O_1 in E_2), the obvious conclusion is that O_1 shows *genetic adaptation* to E_1. The ascription of some form of adaptation to (features of) an organism is obviously a function of the reference organism chosen, that is, of the comparisons we are interested in.

3.1.1. Premise 1. All insectivorous birds in the temperate zone are migrants. Premise 2. Willow warblers are insectivorous birds. Conclusion. Willow warblers are migrants.

3.1.2. The first argument is inductive. The second argument is deductive. We should add the premise that the water in the pond has a low calcium concentration.

3.2. Premise 1. People who have been in close contact with others having an infectious disease, and who are not susceptible, do not get the disease. Premise 2. John has been in close contact with others having an infectious disease, and he is not susceptible. Conclusion. John does not get the disease. This argument is valid. Yet it may not be acceptable. An obvious possibility to define 'susceptible' is as follows. 'Person x is not susceptible to disease y' $=_{df}$ 'x does not get y, not even if x is in close contact with people having y'. According to this definition premise 1 would be unacceptable because it would not have empirical content. If we adopt the definition, we can also reject the argument on the ground that it is a logical circle.

The last part of premise 2 ('John is not susceptable') implicitly contains the conclusion in view of the meaning of 'susceptible'.

3.3.1. This need not be an inadmissible epistemic circle. The conclusion that the plants are infected with a particular virus might be supported by additional, independent evidence.

3.3.2. This is a logical circle. The premise and the conclusion have the same meaning. Notice that logical circles are valid arguments.

3.4.1. In areas with higher temperatures penguins may have more competitors and/or more predators and/or less suitable food sources. The distribution of penguins covaries with many factors in addition to temperature.

3.4.2. It is thought that the devastation of rain forest is indeed one of the causes implicated in carbon dioxide changes. The argument would be valid only if the use of the term 'causes' is meant to imply that no other factors apart from devastation of rain forest are causally implicated (and that effects are not delayed). In fact there are other factors, most notably our burning of fossil fuels. So on this interpretation of 'causes' one of the premises is not true. If 'causes' is taken to refer only to one of the causal factors the premises may be true, but in that case the argument is invalid. Thus the argument at best represents an incomplete explanation.

4.1.1. It is impossible that inconsistent premisses are all *true*.

4.1.2. If the formulation of the argument is taken literally, it appears to be valid since its form is that of a hypothetical syllogism. However, you should notice that the formulation is misleading. The first premise should be taken to mean that there will be no reproduction *as long as* the density is high. Absence of reproduction will of course decrease density to levels which no longer inhibit reproduction. Therefore, if intended meanings are clarified by a reformulation of the argument, it will turn out to be invalid. Also, the premises are false. High density need not in fact inhibit reproduction, and in the absence of reproduction the population could remain in existence due to migration. All in all the argument is totally unacceptable.

4.1.3. Let us use '*p*' for 'lichens are common in the world' and '*q*' for 'worldwide there cannot be much air pollution'. The argument has two premisses, 'if *p* then *q*' and '*p*'; the conclusion is '*q*'. It is valid because it has the form of the modus ponens; this is not to say that it is acceptable. I have taken some vagueness in the statements for granted.

4.2.1. Let '*Lx*' and '*Nx*' stand for '*x* is an area with a soil that has a low nitrogen content' and '*x* is an area without stinging-nettles', respectively. The structure of the explanation is as follows.

$$(x)(\text{if } Lx \text{ then } Nx)$$
$$\frac{La}{Na}$$

4.2.2. No. On the most plausible interpretation (see *3.3*, example *1*), 'prediction' in the argument has two different meanings which should not be covered by the same symbol.

4.3. Inferences without the use of additional information are impossible. However, in view of background knowledge it would be reasonable to infer that the frog will not be present. The detergents are likely to affect the frog in a negative way even in the presence of *Hydrocharis*.

5. Research can only get off the ground if appropriate questions are formulated in the first place. Questions will take some existing theory into account. The biologist will have to provide the questions (of course you can help him with that). He is apparently engaged in an inductivist quest.

6.1. a. Logical statement (unacceptable) with a universal and an existential component. b. Idem. c. Empirical (acceptable) existential statement. d. Possibly a logical statement (depends on the definition of 'species'; if logical, then not acceptable); universal statement. e. Logical (unacceptable) universal statement. f. Empirical (acceptable) universal statement.
Note. Statements can be reconstructed in various ways. The adequacy of a reconstruction will depend on purposes we have (a possible purpose being the evaluation of testability; see the next section). Perhaps you have uncovered an existential component I did not mention. Take f; we could word it as follows: For all *x*, if *x* is a plant which performs photosynthesis, then there will be a substance in *x* which is starch. The statement now appears to have an existential component. I had a different formulation in mind ('... has the property of containing starch').

6.2.1. One assumption is that 'the first radioactive substance we are able to isolate' is indeed 'the first substance which will also show radioactivity'. Also, there are assumptions to the effect that the experimental equipment is in good working order. Further, it is assumed that the stuff injected is indeed '*A*', that the identification of radioactively labelled '*B*' is correct, and that the organism will deal with radioactively labelled *B* and ordinary *B* in similar ways.

6.2.2. It has not been shown that the first test implication is false since there may be unknown sensory organs. The third implication does not follow from the hypothesis. We should rather conclude from the data concerning dead flies, that live flies may achieve orientation through a mechanism which does not essentially involve nervous system activity. The evidence concerning the second implication favors acceptance rather than rejection of the hypothesis.

6.3. a. If 'genetically (environmentally) determined' means 'determined by genetic (environmental) and not by environmental (genetic) factors', then we are dealing with two false hypotheses which do not exhaust the possibilities. If the possibility is left open that environmental (genetic) factors may be involved besides genetic (environmental) ones, then the hypotheses are not exclusive, and trivially true. b. Proper alternatives. c. The alternatives are improper because they are not exhaustive (see *2.4*, example 2). d. The alternatives are not exhaustive.
Note. Alternatives which are not exhaustive from a logical point of view may still be proper, if there is empirical evidence against other possibilities. This may have been the case when researchers pitted the alternatives concerning encephalitis against each other.

6.4.1. Effects of highly diluted substances are implausible according to ordinary physical and chemical theory (they are not actually inconsistent with it). In view of this the hypothesis that such effects exist should only be accepted if the evidence is relatively strong. Positive evidence obtained in a few double-blind experiments will not suffice since such experiments are not foolproof. The decision to accept the hypothesis, or to reject it, would be premature.

6.4.2. This would be reasonable. The 'experiment' in itself is not decisive, but the assumption that insulin is effective in this case is also supported by a well-confirmed biomedical theory.

6.4.3. Can you imagine a surgeon who has no information on the treatment he is giving?

7.2.1. The statement is empirical, and general. It is not universal in the strictest sense of the term because it contains an existential quantifier besides a universal one. It expresses the idea that *there is* a response (nature not specified) ... . For this reason those who have a restrictive view of laws would not regard the statement as a law. To count as a law the statement should also cohere with other generalities, and it should be well-confirmed so that we can assume that it is generally valid. It is unlikely that the statement satisfies the latter criterion. Counter-examples easily come to mind. If you eat food without vitamins you will eventually get ill. The body does not have the resources to counteract all sorts of negative effects.

Perhaps the statement could be qualified so that we get a law for a restricted set of adverse conditions. Notice that it could be regarded as a law under the semantic view. Under this view it characterizes ideal systems. If we adopt this terminology, an important issue is whether we can elaborate valid general, universal theoretical hypotheses concerning empirical systems which exemplify the law.

7.2.2. The authors apparently assume that observations can falsify models (cf. their use of the term 'disprove'). This implies that some hypotheses in models must have a universal component and no existential component. They need not be universal in the philosopher's sense. For example, a model for a particular species is not universal in this sense.

7.3.1. The theory is universal in the sense that it contains universal quantifiers. It is not universal in a different sense: there is explicit reference to particular species. This implies that the theory is not highly general, though it will contain relatively general statements (for example a statement concerning the impact of heterozygote superiority). This is not an example of reduction. We could not formulate an adequate, separate theory for the phenotype level and *then* reduce it to a theory about the genetic level. Ideas about genetic factors are necessary for the proper distinction of patterns at the phenotype level. Likewise, the study of the genes involved will rely on phenomena at the phenotype level right from the start.

7.3.2. The fact that entities at higher levels are composed of lower level entities by no means implies that we can define concepts for higher levels with the help of lower level concepts. So the argument is wrong.

7.3.3. It is conceivable that we can formulate a coherent body of laws which only describe developmental sequences. Hull is here using a criterion of coherence that is much stronger than the one he introduced on pp. 70-71. In fact he seems to demand that a theory for some level, in order to really count as a theory, must cohere with a different *theory*. The 'ordinary' criterion of coherence concerns relationships among statements within one theory. Moreover, Hull assumes that coherence must take the form of reduction (cf. 'derive'). That is surprising since he is a staunch opponent of reductionism. The criterion Hull applies to developmental biology is not acceptable because it is too strong.

7.4.1. Computer science may account for some *aspects* of memory in man. To the extent that the term 'memory' is used for mental aspects of memory, computer science has little to say about it, unless one assumes (as some philosophers do, to my surprise) that computers exhibit mental phenomena.

7.4.2. We *experience* conscious thought as a simple process. It does not follow that a mechanism that makes conscious thought possible has to *be* simple. In fact the opposite is the case. Griffin's argument is based on an erroneous assumption which is easily overlooked.

7.4.3. The core of Bunge's mind-body theory is logically true. It flies in the face of the requirement that theories must have empirical content. His argument that mental predicates are 'coarse and vulgar' at best implies that science should elaborate better *mental* predicates. His second argument is inadequate for the same reason. The third argument is even more outrageous. It *presupposes* that Bunge's theory is right. The fourth argument is clearly irrelevant.

8.1.1. This is an adequate deductive explanation. Its logical form is like the one of the argument in example 1.

8.1.2. The assumption that biological phenomena are the result of physical and chemical *processes* would not imply that physical and chemical *theories* suffice to explain them (cf. comments on reduction in *7.3*). So the fact that we cannot explain a phenomenon through physics and chemistry does not warrant the postulation of a vital force. Moreover, it remains possible that we will be able to explain the phenomenon in this way in the future. Lastly, the nature of the vital force is not specified. The concept of vital force is not sufficiently clear.

8.1.3. The principle of Gause would not permit the inference (by deduction) that the two species will coexist in the area. The force of the explanation is weak.
Note. We could argue that one causal factor responsible for coexistence (presence of other limiting factors) has been identified. This could be regarded as a weak form of explanation; see the next section.

8.2.1. The alternatives are compatible and equally acceptable. Their relevance will depend on the context, that is, on the questions we would like to answer.

8.2.2. The statement does not express a causal relationship since it is logically true.

8.3.1. The explanation is acceptable provided the premises are supported by evidence. It involves a kind of law: 'If changes in conditions in the area of distribution of a species do not permit continued survival, and the species cannot migrate to other areas, then it will go extinct'. This law is close to a logical statement, but I would regard it as an empirical statement which is quite obviously true. (It is logically possible that species become established elsewhere by some miraculous process other than migration.) The most substantial information in the example is best regarded as (valuable) natural history. Notice that the argument involves implicit assumptions (the difference in diversity was not present before the ice ages; Europe could sustain more species, but opportunities for colonization by species which *can* cross mountains have been limited).

8.3.2. If 'function' is defined in etiological terms (such that natural selection is mentioned), the thesis becomes a logical statement.

9.2. a. This is clearly a normative statement. The expression 'clinical justification' refers to values mentioned in the next sentence. b. This statement is likewise normative.
Note. The two statements jointly presuppose that transference of more than three embryo's may lead to multiple pregnancies. That is an empirical matter. We are dealing here with facts and values, but the facts are not expressed by any separate statement.

9.3.1. On the first reconstruction the argument is a naturalistic fallacy. The conclusion is a normative, not an empirical statement. We should add the premise that if animals cannot feel pain, then vivisection on them is allowed. The resulting argument is valid. That does not imply that it is acceptable!

9.3.2. The biologist's statement is not acceptable, but let us assume that it is true for the sake of argument. In that case both statements are possibly acceptable. They are not really incompatible since the meaning of 'altruism' and 'egoism' is different in ethics and biology.

Bibliography*

Connections between philosophy and the life sciences have existed over centuries. They have led to coherent disciplines only in the last two decades. The coherence is marked especially in philosophy of biology. Philosophy of medicine is more pluriform, at times even eclectic.

Important journals are *Biology and Philosophy*, *Theoretical Medicine*, and the *Journal of Medicine and Philosophy*. I will confine myself to a selective review of books, particularly in the Anglo-Saxon tradition. Concerning the philosophy of biology Ruse (1988c) presents a much more extensive bibliography of books and articles.

Unless mentioned otherwise the books mentioned below are at a moderately advanced level.

Various books at the basis of the 'new tradition' in philosophy of biology still merit attention. Beckner (1968, original edition 1959) presented a prize-winning analysis of features that distinguish biology from other sciences. Hull (1974) is clearly a foundational text. It is difficult, though, because of its conciseness. More accessible is Ruse (1973), which has been reprinted, deservedly so. More recent core texts which deserve attention are Rosenberg (1985), and especially Ruse (1988a), Thompson (1989), Darden (1991) Schaffner (1992), Bechtel and Richardson (1992), and Sober (1993). Thompson has applied the semantic view of theories to biology. His book is highly accessible. Darden, Schaffner, and Bechtel and Richardson, have modernized classical approaches in the philosophy of science.

The excellent book by Longino (1990) on the role of values in biology and science in general should be added to this list. Many authors voice strong opinions on this general subject. Longino's is the most balanced account known to me. Concerning the role of values in connection with biology, recent books on two specific subjects, animal liberation ethics and environmental ethics, are reviewed elsewhere (Van der Steen, 1992).

Much of the coherence in philosophy of biology is due to a common emphasis on evolution. In view of this you may wish to consult biology texts on evolution.

---

* Not covered by index.

An important classic is Lewontin (1974), who gives a penetrating discussion of problems with testability in population genetics. The book by Endler (1986) is important but difficult. Futuyma (1986) is more accessible. Excellent general philosophical texts dealing with evolution are Kitcher (1983; a good overview; main subject: criticism of creationism; highly accessible), Sober (1984a), Dupré (1987; contributions by scientists and philosophers), Ruse (1989) and Brandon (1990). Further I recommend the anthology edited by Sober (1984b). Another book by Sober (1988; difficult, for specialists) deals specifically with phylogeny reconstruction. Controversies in evolutionary biology are surveyed by Ruse (1982; accessible). A particularly important controversy concerns the levels at which natural selection operates. Various biologists have allotted central importance to the level of genes (classical texts are Williams, 1966, and Dawkins, 1982), but there are dissenters, for example Eldredge (1985). Levels of selection are also discussed in Lloyd (1988; very difficult; also a defense of the semantic view; excellent source of literature), and Hull (1988; puts science, specifically taxonomy, in an evolutionary context; accessible and provocative). Furthermore the anthology of Brandon and Burian (1984) deals with this subject.

The controversy mentioned above plays a role within modern, neo-Darwinian approaches of evolution. This view is nowadays challenged by various authors, most of them scientists; see for example Pollard (1984), Oyama (1985), Ho and Fox (1988), Goodwin and Saunders (1989), and Goodwin, Sabitani and Webster (1989).

Evolutionary biology is often brought to bear on subjects outside biology proper. Interesting texts are Boyd and Richerson (1985; models for cultural selection), Ruse (1988b; perspectives on homosexuality), Hull (1988; evolutionary account of science). An area which has attracted much attention is human sociobiology. It was introduced as an aside in Wilson's (1985) classic. Later on Wilson (1978) and Lumsden and Wilson (1981) put it at centre stage. This has resulted in an enduring controversy among scientists and philosophers. Various books by philosophers of biology have been devoted to the controversy, for example Rosenberg (1980), Ruse (1979, clearly positive), Kitcher (1985, emphatically negative).

Other extensions of evolutionary theory concern epistemology (Ruse, 1986; moderately positive; see also diverse views in Callebout and Pinxter, 1987, Hahlweg and Hooker, 1989), ethics (Ruse, 1986; moderately positive), language (Lieberman, 1984), psychopathology (Bailey, 1987). Literature on these subjects is often programmatic and/or speculative.

An important philosophical question is whether biology should be concerned with mental processes and with the mind-body problem. If you are interested in this you may want to read Nagel's (1986) general philosophical text, which is accessible and thought-provoking. Likewise for the more recent book by Searle (1992, more difficult). Nagel argues that 'the subjective' cannot wholly be captured by any theory. Searle makes a slightly different point. He argues that science and philosophy should account for subjectivity. They failed to do that in the last few centuries. That's why they are deeply flawed. Also important is the book edited by Marcel and Bisiach (1988), which deals with perplexing problems in the scientific and philosophical study of consciousness.

Within biology, Griffin (1984, accessible), in the last fifteen years, has been promoting a new science called cognitive ethology. He argues that ethology should study mental processes in animals. He has certainly made a point, but some of his arguments are not very convincing from a philosophical point of view. A better case has been made by Rollin (1989, accessible), who also links the issue with ethics. Many other books on the issue have appeared in the last few years. Specifically I recommend the book by the biologists Cheney and Seyfarth (1990) on vervet monkeys. Their perceptive views are based on extensive field work.

In cognitive science, a recently forged interdisciplinary venture involving many sciences, the existence of mental processes in animals is taken for granted. However, biology often gets little attention in this area. The reason is that a view called functionalism has many adherents. Functionalists argue that we should concentrate on functional architecture rather than material implementation in the study of the mental. Hence the important role played by computer metaphors. A lucid general text in cognitive science, which defends a recent view called connectionism, is Bechtel and Abrahamsen (1991). Connectionism is often associated with the study of neural networks, so biology does come in, albeit mostly in an abstract way. Churchland (1986) has emphatically opposed the neglect of biology in cognitive science. Her book is a balanced, integrative account of philosophy and science, not least neurobiology. Like most cognitive scientists, she has strong materialist convictions. Other texts in which biology (unlike philosophy this time) gets its share are Ledoux and Hirst (1986) and Eimas and Galaburda (1990).

It will be obvious by now that interdisciplinary integration is an important issue in the philosophy of biology. A good source concerning this theme is Bechtel (1986). My own view is that the value of integration is now overemphasized (Van der Steen and Thung, 1988).

As I said, the philosophy of medicine is less coherent than the philosophy of biology. This hampers the selection of relevant literature. A well-known text is Pellegrino and Thomasma (1981). I do not recommend it since its philosophical approach is highly eclectic. Some excellent texts, for example Wulff (1976, accessible) and Feinstein (1985), deal with methodology in a broad sense. Wulff, Pederson and Rosenberg (1986, accessible) offer a more general introduction to the philosophy of medicine which I recommend. Other introductions are Jensen (1987) and Reznek (1987). The book by Albert et al. (1988, accessible) is not very adequate since it fails to take recent philosophy of science into account. Van der Steen and Thung (1988; not easy) present a survey and analysis with a very broad scope. Schaffner (1992; not easy), in an important book, deals with both biology and medicine.

It is unfortunate that analysis of science gets much less attention in the philosophy of medicine than in the philosophy of biology. Research in the philosophy of medicine often concentrates on the analysis of highly general notions concerning health and illness, autonomy of patients, paternalism, informed consent, and so forth, as in Culver and Gert (1982). The concepts of health and disease are at centre stage in an anthology edited by Caplan et al. (1981) and also, for example, in Fulford (1989). I do not regard these subjects as unimportant, but the domain covered is very restricted.

In medicine itself, as in biology, philosophical methodology seldom gets attention. In biology, systematic zoology is the most important locus of discussions on methodology. Unfortunately, the scope of discussions has often been limited to Popper's philosophy (see issues of *Systematic Zoology* in the seventies and eighties). In medicine, specifically epidemiology, Popper has also been at centre stage; see for example Rothman (1988).

Medicine as a science and as a practice has many critics today. Some authors have even defended the extreme view that it is a source of disease rather than of health. A classic in which this stance is taken is Illich (1975, popular). He exaggerated, but he certainly had a point. An important problem is that it is very difficult to assess effects of medical treatments (see especially White et al., 1985; difficult). Therefore it should not come as a surprise that there are huge differences between countries even within the Western world in diagnoses and treatments. An important, interesting and highly accessible book by the journalist Payer (1988) highlights these differences. Differences between cultures are studied in a discipline called anthropological medicine and, more specifically, the young discipline of anthropological psychiatry founded by Kleinman (1980; see also Kleinman and Good, 1985; both accessible). Kleinman is a psychiatrist and an anthropologist. I highly recommend his writings. They help one overcome the tendency to see limitations in other cultures, but overlook those in the culture around us. Concerning the role of medicine in our culture, the book of McKeown (1984) also has a healthy, sobering effect. He argues that many alleged achievements of medicine in the past were actually due to a large extent to factors such as better hygiene and food.

The relation between 'biological' and mental processes is even more important for medicine than it is for biology. Or so it should be. Medicine is often characterized by a predominantly biological approach. Such an approach is even defended by many psychiatrists (for example Andreasen, 1984; popular). A much more balanced account of psychiatry is given by Healy (1990, accessible).

I have not reviewed medical ethics since it is a rapidly expanding separate field. There are numerous good (and bad) books in this area, which are easy to locate.

References

Ader, R. 1985. Conditioned immunopharmacological effects in animals: implications for a conditioning program of pharmacotherapy. In L. White, B. Tursky, and G.E. Schwartz, eds., *Placebo: Theory, Research, and Mechanisms*, pp. 306-323. New York: Guilford Press.

Albert, D.A., R. Munson, and M.D. Resnik. 1988. *Reasoning in Medicine, An Introduction to Clinical Inference*. London: Johns Hopkins University Press.

Andreasen, N.C. 1984. *The Broken Brain: The Biological Revolution in Psychiatry*. New York: Harper and Row.

Bailey, B. 1987. *Human Paleopsychology: Applications to Aggression and Pathological Processes*. London: Erlbaum.

Beatty, J. 1980. Optimal-design models and the strategy of model building in evolutionary biology. *Philosophy of Science* 47: 532-561.

Bechtel, W., ed. 1986. *Integrating Scientific Disciplines*. Dordrecht: Nijhoff.

Bechtel, W., and A. Abrahamsen. 1991. *Connectionism and The Mind, An Introduction to Parallel Processing in Networks*. Oxford: Blackwell.

Bechtel, W., and R.C. Richardson. 1992. *Discovering Complexity: Decomposition and Localization as Strategies in Scientific Research*. Princeton: Princeton University Press.

Becker, G., and U. Speck. 1964. Untersuchungen über Magnetfeldorientierung von Dipteren. *Zeitschrift für Vergleichende Physiologie* 49: 301-340.

Beckner, M. 1968. *The Biological Way of Thought*. Berkeley and Los Angeles: University of California Press, second edition.

Bonner, J.T. 1980. *The Evolution of Culture in Animals*. Princeton: Princeton University Press.

Boyd, R., and P.J. Richerson. 1985. *Culture and the Evolutionary Process*. Baltimore: Johns Hopkins University Press.

Brandon, R.N. 1990. *Adaptation and Environment*. Princeton: Princeton University Press.

Brandon, R.N., and R.M. Burian, eds. 1984. *Genes, Organisms, Populations: Controversies over the Units of Selection*. Cambridge Mass.: MIT Press.

Brooks, D.R., and D.A. McLennan. 1991. *Phylogeny, Ecology, and Behavior: A Research Program in Comparative Biology.* Chicago: University of Chicago Press.

Bunge, M. 1980. *The Mind-Body Problem: A Psychobiological Approach.* Oxford: Pergamon.

Byerly, H., and R. Michod. 1991. Fitness and evolutionary explanation. *Biology and Philosophy* 6: 1-22.

Callebout, W., and R. Pinxter (eds) 1987. *Evolutionary Epistemology: A Multiparadigm Program.* Dordrecht: Reidel.

Caplan, A.L., H.T. Engelhardt and J.J. McCartney, eds. 1981. *Concepts of Health and Disease.* London: Addison-Wesley.

Cheney, D.L., and R.M. Seyfarth. 1990. *How Monkeys see the World.* Chicago: University of Chicago Press.

Churchland, P. S. 1986. *Neurophilosophy: Toward a Unified Science of the Mind-Brain.* Cambridge Mass.: MIT Press.

Cossins, A.R. and K. Bowler. 1987. *Temperature Biology of Animals.* London and New York: Chapman and Hall.

Coulter, H.L. 1984. Homeopathy. In J.W. Salmon (ed.). *Alternative Medicines: Popular and Policy Perspectives.* New York: Tavistock Publications, pp. 57-79.

Culver, C., and B. Gert. 1982. *Philosophy in Medicine: Conceptual and Ethical Issues in Medicine and Psychiatry.* New York: Oxford University Press.

Darden, L. 1991. *Theory Change in Science: Strategies from Mendelian Genetics.* New York: Oxford University Press.

Dawkins, R. 1982. *The Extended Phenotype: The Gene as the Unit of Selection.* Oxford: Freeman.

Donovan, S.K., ed. 1989. *Mass Extinctions: Processes and Evidence.* London: Belhaven Press.

Dupré, J., ed. 1987. *The Latest on the Best: Essays on Evolution and Optimality.* Cambridge Mass.: MIT Press.

Dijkgraaf, S., and H.J. Vonk, eds. 1971. *Vergelijkende Dierfysiologie.* Utrecht: Oosthoek.

Ehrlich, P.R., and P.H. Raven. 1969. Differentiation of populations. *Science* 165: 1228-1232.

Eimas, P.D., and A.M. Galaburda. 1990. *Neurobiology of Cognition.* Cambridge Mass.: MIT Press.

Eldredge, N. 1985. *Unfinished Synthesis*. New York: Oxford University Press.

Eldredge, N., and S.J. Gould. 1972. Punctuated equilibria: an alternative to phyletic gradualism. In T.J.M. Schopf, ed., *Models in Paleobiology*, pp. 82-115. San Francisco: Freeman Cooper.

Endler, J.A. 1986. *Natural Selection in the Wild*. Princeton: Princeton University Press.

Engelhardt, H.T. 1981. The concepts of health and disease. In A.L. Caplan, H.T. Engelhardt and J.J. McCartney, eds, *Concepts of Health and Disease*, pp. 31-46. London: Addison-Wesley. Reprint, original article 1975.

Falk, R. 1990. Between beanbag genetics and natural selection. *Biology and Philosophy* 5: 313-325.

Faust, D. 1984. *The Limits of Scientific Reasoning*. Minneapolis: University of Minnesota Press.

Feinstein, A.R. 1985. *Clinical Epidemiology: The Architecture of Clinical Research*. Philadelphia: Saunders.

Fry, J.C., and M.J. Day 1990. *Bacterial Genetics in Natural Environments*. London: Chapman and Hall.

Fulford, K.W.M. 1989. *Moral Theory and Medical Practice*. New York: Cambridge University Press.

Futuyma, D.J. 1986. *Evolutionary Biology*. Sunderland Mass.: Sinauer. Second edition.

Ghiselin, M.T. 1987. Species concepts, individuality, and objectivity. *Biology and Philosophy* 2: 127-143.

Giere, R.N. 1979. *Understanding Scientific Reasoning*. New York: Holt, Reinhart and Winston.

Gifford, F. 1990. Genetic traits. *Biology and Philosophy* 5: 327-347.

Goodwin, B., and P. Saunders, eds. 1989. *Theoretical Biology, Epigenetic and Evolutionary Order from Complex Systems*. Edinburgh: Edinburgh University Press.

Goodwin, B., A. Sibatani, and G. Webster, eds. 1989. *Dynamic Structures in Biology*. Edinburgh: Edinburgh University Press.

Goossens, W.K. 1980. Values, health, and medicine. *Philosophy of Science* 47: 100-115.

Goudge, T.A. 1961. *The Ascent of Life*. Toronto: University of Toronto Press.

Gould, S.J., and R.C. Lewontin. 1979. The spandrels of San Marco and the Panglossian paradigm: a critique of the adaptationist programme. *Proceedings of the Royal Society London* B205: 581-598.

Griffin, D.R. 1984. *Animal Thinking*. Cambridge Mass.: Harvard University Press.

Grime, J.P. 1979. *Plant Strategies and Vegetation Processes*. Chichester: Wiley.

Haccou, P., and W.J. van der Steen. 1992. Methodological problems in evolutionary biology. IX. The testability of optimal foraging theory. *Acta Biotheoretica* 40: 285-295.

Hahlweg, K., and C.A. Hooker, eds. 1989. *Issues in Evolutionary Epistemology*. Albany: SUNY Press.

Hairston, N.G., F.E. Smith and L.B. Slobodkin 1960. Community structure, population control and competition. *American Naturalist* 94: 421-425.

Harding, S. 1976. *Can Theories be Refuted?* Dordrecht: Reidel.

Harper, C.W. 1975. Origin of species in geologic time: Alternatives to the Eldredge-Gould model. *Science* 190: 47-48.

Hayflick, L. 1987. Origins of longevity. In Warner, H.R., R.N. Butler, R.L. Sprott, and E.L. Schneider, eds, pp. 21-34. *Modern Biological Theories of Aging*. New York: Raven Pres.

Healy, D. 1990. *The Suspended Revolution: Psychiatry and Psychotherapy Re-examined.* London: Faber and Faber.

Hempel, C.G., and P. Oppenheim. 1948. Studies in the logic of explanation. *Philosophy of Science* 15: 135-175.

Ho, M.-W. and S.W. Fox. 1988. *Evolutionary Processes and Metaphors*. Chichester: Wiley.

Hull, D.L. 1965. The effect of essentialism on taxonomy, two thousand years of stasis, I and II. *British Journal for the Philosophy of Science* 15: 314-326 and 16, 1-18.

_____. 1968. The operational imperative: sense and nonsense in operationism. *Systematic Zoology* 17: 438-457.

_____. 1974. *Philosophy of Biological Science*. Englewood Cliffs, N.J.: Prentice-Hall.

_____. 1978. A matter of individuality. *Philosophy of Science* 45: 335-360.

_____. 1981. Reduction in genetics. *Journal of Medicine and Philosophy* 6: 125-143.

_____. 1988. *Science as a Process: An Evolutionary Account of the Social and Conceptual Development of Science.* Chicago: University of Chicago Press.

Humphries, G.W., and M.J. Riddoch. 1987. *To See but not to See: A Case Study of Visual Agnosia.* London: Erlbaum.

Illich, I. 1975. *Medical Nemesis: The Expropriation of Health.* London: Boyars.

Jansen, M.J.W., and W.J. van der Steen. 1975. Mathematics and methodology in experiments on density effects, illustrated by studies on the freshwater snail *Lymnaea stagnalis* (L.). *Proceedings Koninklijke Nederlandse Akademie van Wetenschappen* C78: 198-205, 1975.

Jensen, U.J. 1987. *Practice and Progress: A Theory for the Modern Health-Care System.* Oxford: Blackwell.

Kitcher, P. 1983. *Abusing Science.* Cambridge Mass.: MIT Press.

_____. 1985. *Vaulting Ambition.* Cambridge Mass.: MIT Press.

Kitcher, P., and W.C. Salmon, eds. 1989. *Scientific Explanation.* Minneapolis: University of Minnesota Press.

Kleinman, A. 1980. *Patients and Healers in the Context of Culture: An Exploration of the Borderland between Anthropology, Medicine and Psychiatry.* Berkeley: University of California Press.

Kleinman, A., and B. Good. 1985. *Culture and Depression: Studies in the Anthropology and Cross-Cultural Psychiatry of Affect and Disorder.* Berkeley: University of California Press.

Ledoux, J.E., and W. Hirst. 1986. *Mind and Brain: Dialogues in Cognitive Neuroscience.* Cambridge: Cambridge University Press.

Lewontin, R.C., 1974. *The Genetic Basis of Evolutionary Change.* New York: Columbia University Press.

Lieberman, P. 1984. *The Biology and Evolution of Language.* Cambridge Mass.: Harvard University Press.

Lipton, P. 1991. *Inference to the Best Explanation.* London: Routledge.

Lloyd, E.A. 1988. *The Structure and Confirmation of Evolutionary Theory.* Westport: Greenwood Press.

Longino, H.E. 1990. *Science as Social Knowledge: Values and Objectivity in Scientific Inquiry.* Princeton: Princeton University Press.

Lovelock, J.E. 1979. *Gaia: A New Look at Life on Earth.* Oxford and New York: Oxford University Press.

_____. 1988. *The Ages of Gaia: A Biography of our Living Earth.* New York and London: Norton and Company.

Lumsden, C., and E.O. Wilson. 1981. *Genes, Mind, and Culture.* Cambridge Mass.: Harvard University Press.

Marcel, A.J., and E. Bisiach, eds. 1988. *Consciousness in Contemporary Science.* Oxford: Clarendon Press.

Mayr, E. 1970. *Populations, Species and Evolution.* Cambridge Mass.: Harvard University Press.

_____. 1987. The ontological status of species: scientific progress and philosophical terminology. *Biology and Philosophy* 2: 145-166.

McKeown, T. 1984. *The Role of Medicine.* Oxford: Blackwell. Second edition.

Mikhail, A. 1985. Stress, a psychophysiological conception. In A. Monat and R.S. Lazarus, eds, *Stress and Coping: An Anthology*, pp. 30-39. New York: Columbia University Press.

Nagel, E. 1961. *The Structure of Science.* London: Routledge.

Nagel, T. 1986. *The View from Nowhere.* Oxford: Oxford University Press.

Oyama, S. 1985. *The Ontogeny of Information, Developmental Systems and Evolution.* Cambridge: Cambridge University Press.

Paul, G.L. 1985. Can pregnancy be a placebo effect? Terminology, designs, and conclusions in the study of psychosocial and pharmacological treatments of behavioural disorders. In L. White, B. Tursky, and G.E. Schwartz, eds, *Placebo: Theory, Research and Mechanisms*, pp. 137-166. New York: Guilford Press.

Payer, L. 1988. *Medicine and Culture: Varieties of Treatment in the United States, England, West Germany, and France.* New York: Holt and Company.

Pellegrino, E., and D. Thomasma. 1981. *A Philosophical Basis of Medical Practice.* Oxford: Oxford University Press.

Pimm, S.L. 1991. *The Balance of Nature: Ecological Issues in the Conservation of Species and Communities.* Chicago: University of Chicago Press.

Pollard, J., ed. 1984. *Evolutionary Theory: Paths into the Future.* Chichester: Wiley.

Popper, K.R. 1959. *The Logic of Scientific Discovery.* London: Hutchinson English translation. Original German edition 1934.

_____. 1963. *Conjectures and Refutations: The Growth of Scientific Knowledge.* New York: Harper and Row.

Quine, W.V.O. 1953. *From a Logical Point of View.* Cambridge Mass.: MIT Press

_____. 1960. *Word and Object*. Cambridge Mass.: MIT Press.

Rapoport, E.H., and O. Rapoport. 1960. Elementary biological functions and the concept of living matter. *Acta Biotheoretica* 13: 1-28.

Raup, D.M. 1991. *Extinction: Bad Genes or Bad Luck?* New York and London: W.W. Norton.

Reilly, D.T., C. McSharry, M.A. Taylor, and T. Atchinson 1986. Is homeopathy a placebo response? Controlled trial of homeopathic potency, with pollen in hayfever as a model. *Lancet* 18 October: 881-886.

Reznek, L. 1987. *The Nature of Disease*. London: Routledge and Kegan Paul.

Richards, R. 1986. A defense of evolutionary ethics. *Biology and Philosophy* 1: 265-293.

Rollin, B. 1989. *The Unheeded Cry: Animal Consciousness, Animal Pain and Science*. Oxford: Oxford University Press.

Rose, M.R. 1991. *Evolutionary Biology of Aging*. New York and Oxford: Oxford University Press.

Rosenberg, A. 1980. *Sociobiology and the Preemption of Social Science*. Baltimore: Johns Hopkins University Press.

_____. 1985. *The Structure of Biological Science*. Cambridge: Cambridge University Press.

Rothman, K.J., ed. 1988. *Causal Inference*. Chestnut Hill: Epidemiology Resources.

Ruse, M. 1973. *The Philosophy of Biology*. London: Hutchinson.

_____. 1979. *Sociobiology: Sense or Nonsense?* Dordrecht: Reidel. Second edition 1985.

_____. 1982. *Darwinism Defended: A Guide to the Evolution Controversies*. Reading: Addison-Wesley.

_____. 1986. *Taking Darwin Seriously: A Naturalistic Approach to Philosophy*. Oxford: Blackwell.

_____, ed. 1988a. *What the Philosophy of Biology is*. Dordrecht: Kluwer.

_____. 1988b. *Homosexuality: A Philosophical Inquiry*. Oxford: Blackwell.

_____. 1988c. *Philosophy of Biology Today*. Albany: SUNY Press.

_____. 1989. *Molecules to Men: The Concept of Progress in Evolutionary Biology*. Cambridge Mass.: Harvard University Press.

Salmon, W.C. 1989. *Four Decades of Scientific Explanation*. Minneapolis: University of Minnesota Press. Reprint from Kitcher and Salmon 1989.

Schaffner, K.F. 1967. Approaches to reduction. *Philosophy of Science* 34: 137-147.

_____. 1980. Theory structure in the biomedical sciences. *The Journal of Medicine and Philosophy* 5: 57-97.

_____. 1986. Exemplar reasoning about biological models and diseases: a relation between the philosophy of medicine and philosophy of science. *The Journal of Medicine and Philosophy* 11: 63-80.

_____. 1992. *Discovery and Explanation in Biology and Medicine.* Chicago: University of Chicago Press.

Schwartz, S.R., and B. Africa. 1984. Schizophrenia disorders. In H.H. Goldman, eds, *Reviews of General Psychiatry*, pp. 311-327. Los Altos: Lange Medical Publications.

Searle, J. 1992. *The Rediscovery of the Mind.* Cambridge Mass.: MIT Press.

Selye, H. 1983. The stress concept: past, present and future. In C.L. Cooper, ed., *Stress Research: Issues for the Eighties*, pp. 1-20. Chichester: Wiley.

Shrader-Frechette, K.S. 1985. *Science Policy, Ethics, and Economic Methodology.* Dordrecht: Reidel.

Simpson, G.G. 1964. *This View of Life.* New York: Harcourt, Brace and World.

Sloep, P.B., and W.J. van der Steen 1987. The nature of evolutionary theory: the semantic challenge. *Biology and Philosophy* 2: 1-15.

_____. 1988. A natural alliance of teaching and philosophy of science. *Educational Philosophy and Theory* 20: 24-32.

Smart, J.J.C. 1963. *Philosophy and Scientific Realism.* London: Routledge and Kegan Paul.

Sober, E. 1984a. *The Nature of Selection.* Cambridge Mass.: MIT Press.

_____, ed. 1984b. *Conceptual Issues in Evolutionary Biology.* Cambridge Mass.: MIT Press.

_____. 1988. *Reconstructing the Past: Parsimony, Evolution, and Inference.* Cambridge Mass.: MIT Press

_____. 1993. *Philosophy of Biology.* Boulder, Colorado: Westview Press.

Sokal, R.R., and T.J. Crovello. 1970. The biological species concept: a critical evaluation. *American Naturalist* 104: 127-153.

Stephens, D.W., and J.R. Krebs. 1986. *Foraging Theory.* Princeton: Princeton University Press.

Stanley, S. 1987. *Extinction.* San Francisco: Freeman.

Thompson, P. 1989. *The Structure of Biological Theories*. Albany: SUNY Press.

Toates, F. 1986. *Motivational Systems*. Cambridge: Cambridge University Press.

Van der Steen, W.J. 1967. The influence of environmental factors on the oviposition of *Lymnaea stagnalis* (L.) under laboratory conditions. *Archives Néerlandaises de Zoologie* 17: 403-468.

_____. 1981. Testability and temperature adaptation. *Oikos* 37: 123-125.

_____. 1990a. Concepts of biology, a survey of practical principles. *Journal of Theoretical Biology* 143: 383-403.

_____. 1990b. Interdisciplinary integration in biology? *Acta Biotheoretica* 38: 23-36.

_____. 1991. *Denken over Geneeskunde, een Praktische Filosofie voor de Gezondheidszorg*. Lochem: De Tijdstroom.

_____. 1992. Ethics, animals and the environment: a review of recent books. *Acta Biotheoretica* 40: 339-347.

_____. 1993a. Towards disciplinary disintegration in biology. *Biology and Philosophy*.

_____. 1993b. Towards a practicable methodology for medicine: the impact of conceptual analysis, *Perspectives in Biology and Medicine*.

Van der Steen, W.J., and H. Kamminga. 1991. Laws and natural history in biology. *British Journal for the Philosophy of Science* 42: 445-467.

Van der Steen, W.J., and A.W. Musschenga. 1992. The issue of generality in ethics. *Journal of Value Inquiry* 26: 511-524.

Van der Steen, W.J., and M. Scholten. 1985. Methodogical problems in evolutionary biology. IV. Stress and stress tolerance, an exercise in definitions. *Acta Biotheoretica* 34: 81-90.

Van der Steen, W.J. and P.B. Sloep. In press. Philosophy, education, and the explosion of knowledge. *Interchange*.

Van der Steen, W.J., and P.J. Thung. 1988. *Faces of Medicine: A Philosophical Study*. Dordrecht: Kluwer.

Van Fraassen, B.C. 1980. *The Scientific Image*. Oxford: Clarendon Press.

Voorzanger, B. 1984. Altruism in sociobiology, a conceptual analysis. *Journal of Human Evolution* 13: 33-39.

_____. 1987. *Woorden, Waarden, en de Evolutie van Gedrag: Humane Sociobiologie in Methodologisch Perspectief*. Amsterdam: Free University Press.

Warner, H.R., R.N. Butler, R.L. Sprott, and E.L. Schneider, eds. 1987. *Modern Biological Theories of Aging*. New York: Raven Press.

Weiskrantz, L. 1987. *Blind Sight: a Case Study and its Implications*. Oxford: Oxford University Press.

Weiskrantz, L., E.K. Warrington, M.D. Sanders, and J. Marshall. 1974. Visual capacity in the hemianopic field following a restricted occipital ablation. *Brain* 97: 709-728.

White, L., B.Tursky, and G.E. Schwartz, eds. 1985. *Placebo. Theory, Research, and Mechanisms*. New York: Guilford Press.

Williams, G.C. 1966. *Adaptation and Natural Selection: A Critique of Some Current Evolutionary Thought*. Princeton: Princeton University Press.

Wilson, E.O. 1978. *On Human Nature*. Cambridge Mass.: Harvard University Press.

_____. 1985. *Sociobiology: The New Synthesis*. Cambridge Mass.: Harvard University Press.

Wulff, H.R. 1976. *Rational Diagnosis and Treatment*. Oxford: Blackwell.

Wulff, H.R., S.A. Pederson and R. Rosenberg. 1986. *Philosophy of Medicine: An Introduction*. Oxford: Blackwell.

Index

abundance, 11, 25,45, 46, 65, 81, 108, 110,
acceptability, 7, 36, 152
acclimation, 105
accompanying feature, 13
acid rain, 35
ad hoc, 89, 90, 112
adaptation, 30, 42
adaptation; capacity —, 105; genetic —,
    105; non-genetic, —, 105; physiological
    —, 112; resistance — 105, temperature
    —, 104, 105, 109
Ader, R., 98
Africa, B., 28
aging, 1-4
AIDS, 57,58, 66
air pressure, influence of — , 79-80
alarm call, 137,138
allele, 136
alternative explanations, 137
alternative hypotheses, 92-94
alternative hypothesis (statistics), 92
altruism, 18, 19, 22, 25, 119, 153, 155
ambiguity, 10, 11, 40, 41, 116, 119
analytic statement, 6
antecedent, 50
antibiotic, 130.131, 152, 153
antibody, 16
argument, 3, 4, 10, 33-46; circular —, 37,
    39, 42, 43, 131; deductive —, 33-43;
    inductive —, 33-36, 44-46; probabilistic
    —, 61-63, 134-136; validity of —, 3, 4,
    36, 52, 54, 70, 126-128
argumentum ad consequentiam, 42, 152, 153
argumentum ad hominem, 41
argumentum ad ignorantiam, 41, 42
arsenic poisoning, 33, 34
artefact, 88
arteriosclerosis, 16
arthritis, 16
assumption or presupposition, 11, 87-89,
    90, 107
auto-immune disease, 98
autonomy of biology, 102

background information or theory, 45, 58,
    60, 95, 97
Bacon, F., 76
Beatty, J., 108, 110
Becker, G., 91
Beckner, M., 20

bias, 91, 92, 156, 157, 173
biogenerative system, 19
biological species concept, 12, 13, 22-24
birth rate, 106
Bonner, J.T., 119
Bowler, K., 105
Brooks, D.R., 145
Bunge, M., 122, 123
Byerly, H., 82

camouflage, 135, 139
carnivore, 25, 110
causation, 35, 44-46, 63, 64-66, 101, 102,
    125, 126, 132-139, 143, 144, 146, 157
chlorophyll, 128
circadian rhythm, 14, 41
circle in definition, see definition
circular argument, 37, 39, 42, 43, 131
clarification or explication by definition, 17
clarity, 5, 10-12, 31, 80, 81, 103
classification; — and concepts 24, 25, 75; —
    in biology 19, 20; —in medicine, 28,
    29, 157; exclusive —, 25, 92;
    exhaustive —, 25, 92-94, 137; — of
    hypotheses, 92-94; — of plant com-
    munities, 178
closed system, 51
cluster concept, 20
coexistence of competitors, 110, 132
cognitive statement, 7
coherence, 21, 23, 38, 39, 44, 45, 47, 63,
    66, 67, 83, 102, 103, 113-117, 129,
    130
comparative concept, 14
competition 12, 109, 138
complexity of organisms, 136
complexity, ecological —, 81
compound statement, 49
concept, 2-3, 9-32; cluster —, 20; coherence
    of —, 21; comparative, —, 14; —
    loaded with theory, 15, 29, 31, 75, 119-
    121; complex —, 27-30; empirical
    reference of —, 21-22; normative, —
    16; observational —, 14; operationality
    of —, 21-22; polythetic —, 20;
    primitive —, 26; qualitative, — 14;
    quantitative —, 14; theoretical —, 14;
    theoretical significance of —, 23, 24
conceptual analysis, 3, 26-30, 118-121
conditioning, 98

203